CONCRETE
WORK
SIMPLIFIED

Donald R. Brann

Library of Congress Card No. 66-24876

TWELFTH PRINTING —1976
REVISED EDITION

Published by
DIRECTIONS SIMPLIFIED, INC.

Division of
EASI-BILD PATTERN CO., INC.
Briarcliff Manor, NY 10510

FIRST PRINTING
©1960

REVISED EDITIONS
1966,1969,1971,1973.
1974,1976

ISBN 0-87733-617-2

PUT WORRY TO WORK

If we recognize that God's gift to each of us consists of time and energy, and this inheritance is ours to use as we please, we realize why some people are luckier than others, why some win fame and fortune, while others can't get off first base.

Time and energy is a currency that's immune to inflation and taxation. It's one of the easiest to squander, and when spent, can't be regained. Think of this the next time you switch on TV and spend countless hours dialing around a wasteland of nonsense, or spend equal time worrying about a problem only time can resolve.

Worrying is like spinning your wheels in mud or snow. It's a mental habit we cultivate early in life that grows and wastes more of our inheritance than we realize. We don't suggest breaking this habit, we suggest harnessing it.

Remember — every thought you conceive, every move you make is fueled by particles of time and energy in the same way gasoline powers an engine. Invest spare time making repairs or improvements, and you'll discover how to double the purchasing power of a tax shrunken buck.

This book tells how to work with concrete, how to invest time improving your home and your well being. Just as every house requires continual care, so your mind requires continual release from tension. Working with concrete provides an economical solution to costly jobs while it provides needed physical exercise and mental relief.

Make this test. The next time a problem starts your worry machine, reach, read and start working on a home improvement job. Note how quickly you find relief when you refocus your Mentalens.

TABLE OF CONTENTS

IT'S EASY WHEN YOU KNOW WHAT HAS TO BE DONE

Concrete work isn't difficult nor does it take a great deal of skill. It does require knowing how, and a willingness to do. This book describes all the different kinds of work a homeowner may find necessary. Each is described and illustrated. Since many jobs require knowing many different steps, read the book through completely so you can refer to procedures pertinent to the work you need to do.

CONCRETE MIX

Making concrete requires drinking-quality water, clean sand, gravel or crushed stone. Use sand and gravel free of silt, clay and loam.

For small jobs, use the premixes. These are exactly what the name implies, an exact mixture. Add water as specified and they are ready to use. A gravel mix contains cement, sand and gravel. The gravel mix is used for concrete 2″ or more thick. Ideal for setting posts, fixing large holes. A mortar mix contains cement, lime and fine sand. A sand mix, cement and sand is used for patching cracks, plastering, parging or stuccoing. 80 lbs of premix will cover about 16 sq. ft. ¼″ thick when used for plastering. The mortar mix is perfect for laying brick or blocks. Also used to repair mortar joints in chimneys, etc.. Premixed concrete only requires adding water. Mix well, then use. Premixes* are not recommended for large jobs.

Illus. 1 indicates mixtures recommended for use specified. For large jobs buy ready-mix. For medium size jobs, or where you can't get a ready-mix truck close enough to the work area, mix your own. Rent a mixing machine. These are great labor savers and well worth their cost.

Small amounts of concrete can be mixed in a steel wheelbarrow. Use a straight sided can or pail as a measure. One part cement, 2¼ parts of sand, 3 parts of gravel provide what is known as a 1-2-3 mix. This is an ideal mix for most improvement work.

*Note exception on page 92.

(1)

Where Used	Cubic Ft. of Sand	Cubic Ft. Gravel	Bag of Cement	Approx. Gals. of Water
Floors, steps, walks over 2" thick	2¼	3	1	5
Footings, foundation walls, retaining walls	2¾	4	1	5½
Two course floors, pavements, flower boxes, benches, and where concrete is 1" to 2" thick	1¾	2¼	1	4

To estimate amount of concrete needed for various size areas, multiply the width by length to obtain square footage, then note chart. Illus. 2.

ESTIMATING MATERIALS

The table below gives the number of cubic yards of concrete in slabs of different thicknesses and areas.

Multiply the slab length by width to get square feet. Then read quantity of concrete from the table for desired thickness.

EXAMPLE: The slab is 20x30 ft. and 4 in. thick.

Area = 20 × 30 = sq. ft.

Since the table does not go as high as 600 sq. ft., use the concrete quantity for 300 sq. ft. and multiply it by 2.

Quantity for 300 sq. ft. = 3.7 cu. yd.
2 × 3.7 = 7.4 cu. yd.

CUBIC YARDS OF CONCRETE IN SLABS

(2)

Area in square feet (length × width)	Thickness in inches				
	4	5	6	8	12
50	0.62	0.77	0.93	1.2	1.9
100	1.2	1.5	1.9	2.5	3.7
200	2.5	3.1	3.7	4.9	7.4
300	3.7	4.7	5.6	7.4	11.1
400	4.9	6.2	7.4	9.8	14.8
500	6.2	7.2	9.3	12.4	18.6

Use as little water as possible when mixing concrete. When mixing mortar, use as much water as the mortar will take.

MORTAR MIX

Mortar is a mixture of cement and sand. Or cement, hydrated lime and sand. Mortar is used when laying brick, block or stone.

Initially, practice laying block or brick with prepared mortar mixes. These only require adding water in exact proportions directions on bag recommend. Dump the contents in a wheelbarrow, spread it out, make a hole in the middle and add as much water as directions specify. Use a hoe to pull the dry mix into the water. Or use masonry cement. One part masonry cement to two to three parts mortar sand is an acceptable mix.

Codes frequently specify using a Type M or N mortar mix. This is what each contains.

Type M — 1 part Portland cement, ¼ part hydrated lime and not less than 2¼ or more than 3 parts sand by volume. This is a high strength mortar and is suitable for reinforced brick or block masonry, or other masonry below grade that is in contact with the earth, such as foundations, retaining walls, walks, sewers, manholes, catch basins.

Type N — 1 part Portland cement, 1 part hydrated lime and not less than 4½ or more than 6 parts sand by volume. This mortar is considered a medium strength mortar and is recommended for exposed masonry above grade, walls, chimneys, and exterior brick and block work subject to severe exposure.

Always use a batch of mortar within 2½ hours after it has been mixed. Always add water when needed. Always keep it alive by turning it over with a trowel.

MEASURING BOX

To accurately measure ingredients, build a bottomless box, Illus. 3. Cut two pieces of ½", ⅝" or ¾" plyscord, A — 13½"x12"; two pieces B — 12"x12". Apply waterproof glue and nail A to B using 8 penny common nails spaced three inches apart. Cut two 1x2x24" for handles. Nail handles to box in position shown. Use a file or rasp to round edges of handles. When filled level with top, measure holds one cubic foot.

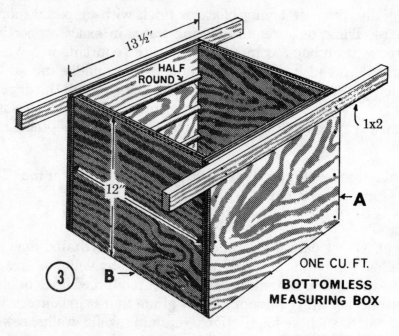

To simplify measuring smaller quantities, nail strips of ⅜" half-round to inside of box. For a quarter cubic foot, nail one strip three inches from botton; nail another six inches from bottom for a half cubic foot; nine inches from bottom for three quarters of a cubic foot. Always place the bottomless measure in mortar tub. When you fill the amount required, remove measure.

A bag of Portland Cement weighs 94 lbs.* It is equal to one cubic foot. One bag of cement, 2¼ cubic feet of sand, plus 3 cubic feet of gravel and between 4 to 5½ gallons of water makes approximately 4½ cubic feet of concrete.

*87.5 lbs. in Canada.

CONCRETE MIXING TUB

You can mix concrete in an oversized wheelbarrow or build a mixing tub, Illus. 4. Using 1x12, cut two sides A, to angle shown, and two ends*B. Apply waterproof glue and nail sides to ends. Apply glue and nail a ³⁄₁₆″ tempered hardboard panel to bottom. Always place tub on a level surface. If you use it where it needs to be supported to make level, nail an extra ⅜″ or ½″ plywood panel to bottom as a stiffener.

*Plane top and bottom edge to angle required.

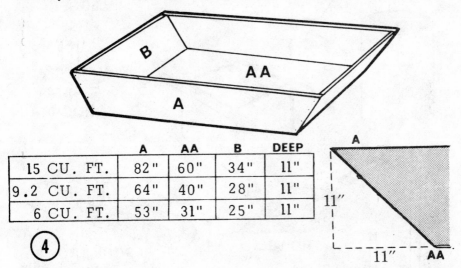

	A	AA	B	DEEP
15 CU. FT.	82″	60″	34″	11″
9.2 CU. FT.	64″	40″	28″	11″
6 CU. FT.	53″	31″	25″	11″

④

Paint tub with wood preservative. When thoroughly dry, paint inside surface with used crankcase oil. After using, always scrape out concrete and hose tub thoroughly. When dry, paint with old crankcase oil before reusing.

TOOLS REQUIRED

Psychologists claim the more physical work one does, and gets accustomed to doing, the greater will be their measure of peace of mind. They advise purchasing tools for concrete work with the same thought as sporting equipment. Just as no one attempts playing any game until they have the necessary equipment, concrete also requires proper tools, Illus. 5. These consist of a shovel, mason's hoe, measuring box, wheelbarrow, mixing tub, trowel, brick hammer, chisel, four foot level, a ball of non-stretching nylon line or a chalk line, a wood and/or steel float, edger, groover, 6 ft. folding rule, etc. You will also need a 5 gallon can or 50 gallon drum to wash off tools, or as a reservoir for water mixed with additive or anti-freeze.

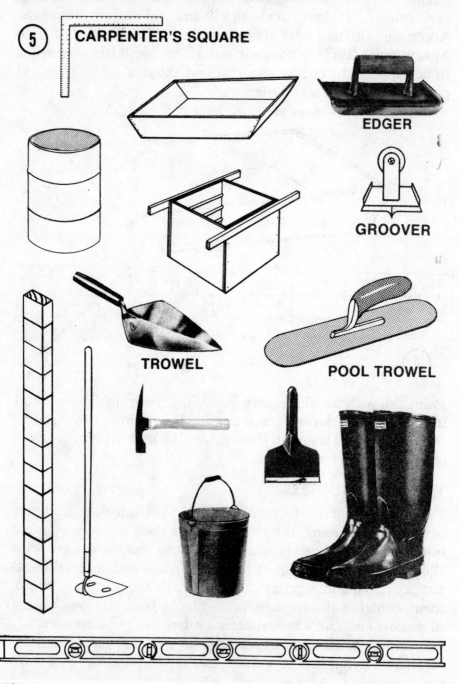

(5) CARPENTER'S SQUARE

EDGER

GROOVER

TROWEL

POOL TROWEL

Get acquainted with your tool rental store. They usually have all the tools the pros use. These include a long handled bull float, roller tamper, magnesium screed, and many other labor-saving tools.

Keep tools store fresh. Always wash tools after using. Apply a light coating of oil to prevent rust. Concrete tools that are mis-used won't permit you to do the best kind of work.

BASIC FACTS

Always store Portland cement in a dry place. Only buy as many bags as you can use during the period you are working. Never lay bags on the ground or on a concrete floor, always on boards placed across blocks. Always cover with polyethelyne when not being used. Dampness, morning dew, even humidity, can harden cement.

A mixing tub, Illus. 4, provides the best way to mix a full bag of cement. A steel wheelbarrow can be used for smaller batches. Always use up a batch of concrete as quickly as possible. Prefer-ably within 30 to 40 minutes. During hot weather, in less time.

The secret to good concrete lies in accurately measuring all materials, mixing the sand and cement thoroughly to one con-sistent color before adding gravel. Mixing gravel, sand and cement thoroughly before adding water. Use only as much water as is needed to make a consistency that holds its shape when a handful is compressed. The exact amount of water is difficult to specify since the moisture content of sand varies.

CURING

Concrete requires time to cure. During this time wet concrete with a fine mist. If this can be done once or twice every 24 hours for a least three days, it helps the curing process. In a basement or garage, or other exposed slab, spray a fine mist. If job is ex-posed to direct sun, cover concrete with burlap, roofing felt or building paper. Remove same before wetting concrete.

Note: Don't attempt to do a big concrete job early on a hot day. Concrete sets up fast in direct sunshine. Better wait until mid-afternoon, even if it means working late in the cool of evening.

FORMS

Concrete work requires forms, Illus. 6. These can be 2x4, 2x6, ⅝″ or ¾″ plyscord. Use plyscord when building large forms; or a form over an outcropping of rock, Illus. 7; or when pouring a concrete foundation or reinforced retaining wall, Illus. 8. Rental forms are also available. These are easy to assemble and save time and material needed to build your own.

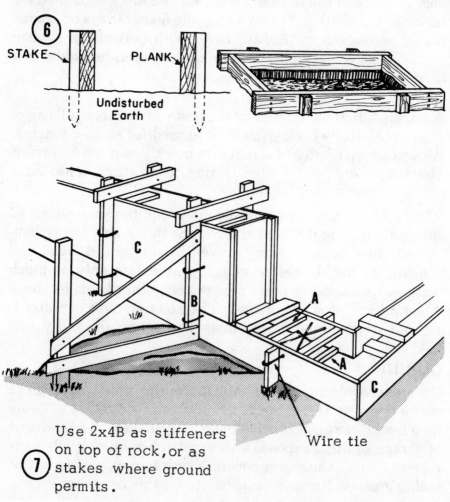

STAKE   PLANK

Undisturbed Earth

Use 2x4B as stiffeners on top of rock, or as stakes where ground permits.

Wire tie

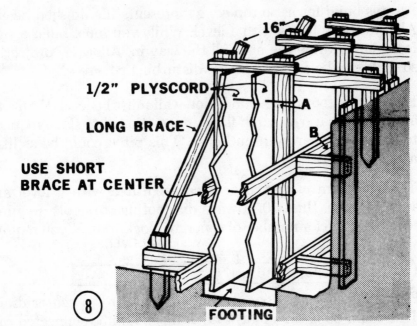

16"

1/2" PLYSCORD

LONG BRACE

USE SHORT
BRACE AT CENTER

A

B

FOOTING

⑧

When working on a large job, such as a floor, footing, etc., where you can't finish the entire job in one pouring, divide the area with forms. Build sections to a size you can handle, Illus. 9. A 4' to 6' wide form is an easy size for two men to screed.

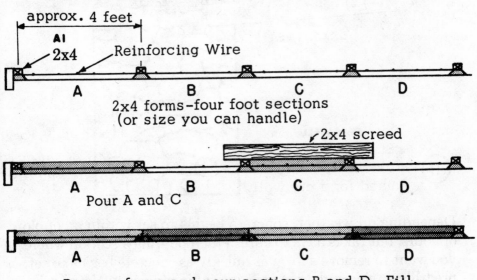

approx. 4 feet

Al
2x4

Reinforcing Wire

A   B   C   D

2x4 forms-four foot sections
(or size you can handle)

2x4 screed

A   B   C   D

Pour A and C

A   B   C   D

⑨ Remove forms and pour sections B and D. Fill in area between Al and form. Second pouring is indicated by darker shading.

Always build forms so top edge represents the finished height. Always check forms with a level. While you must build a rigid form, never drive the nails all the way in. Allow ½" projection so nails can be pulled without disturbing concrete.

Forms for a footing or foundation wall must be level. Forms for a sidewalk, patio or porch floor are positioned to slope required to drain water away from house. This pitch could be as little as ⅛" to a foot.

Check bottom of trench or excavated area with a level and straight edge, Illus. 10. Always dig footings to depth required to allow space for number of concrete block courses you require.

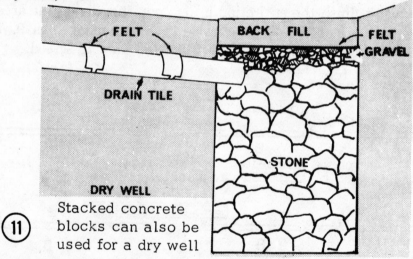

(10) 2x6x8'or10'or12' ──LEVEL

Forms for a garage floor should slope towards a center drain. This could be ⅛" to a foot. The drain should be connected to a dry well, Illus. 11.

FELT   BACK FILL   FELT
GRAVEL
DRAIN TILE
STONE
DRY WELL
(11) Stacked concrete blocks can also be used for a dry well

Depending on weather, concrete begins to set within a couple of hours, but takes days to cure. When pouring a section where you want to remove a form as quickly as concrete begins to set, predrill one side and drive nails into adjoining part. Never drive nails all the way. This simplifies removal of form without disturbing concrete.

When doing a large area in sections, always wet edge of existing concrete before pouring next batch. To make a better bond, paint edge with a wet mix consisting of 1 part cement to 2 parts of sand. This acts as a bonding joint.

The site selected for any concrete work must be cleared of all foreign matter. All sod, shrubbery, stumps, etc. must be removed. Large boulders should also be removed. Frost can get under a large boulder and crack a footing or foundation wall. If site selected has any soft spots, mud, humus, etc., it must be removed and holes filled with gravel. Always tamp gravel to make certain it's compacted to provide an equal bearing.

If base is dry and hard and there's a natural runoff, you can spray area before laying concrete. Only spray, don't make puddles or mud. If area is questionable, build it up with gravel and extend gravel base at least a foot beyond building site. A 6″ buildup of gravel is not too much.

Straight lengths of 1x6, 2x4, plyscord, or other dimension lumber may be used for forms. Always raise one end of a board eye level and "sight down." Note whether the edge is straight, raised, or contains a low spot. Use only straight edged lumber as the top edge of the form helps shape the surface of the concrete.

Always excavate area for a footing form, Illus. 6., to a depth below frost level. In frost free areas, form can be placed on any dry, compacted area that's scraped clean of greenery, stumps or growth of any kind. Never build a form or lay concrete over a recently filled area. Never excavate, then throw dirt back to level area. If you find it necessary to fill in, use gravel. Always tamp gravel when used to back fill.

If site is compacted but wet, remove soft spots, spread and tamp gravel. Never lay concrete in mud.

Use 2x4's sharpened at end for form stakes, or rent iron stakes if you plan a large form. Only toenail forms to stakes so you can remove forms without disturbing concrete.

Where you can't drive stakes, use wire spreaders, Illus. 7,12, or 1 x 2's to spread and hold form together.

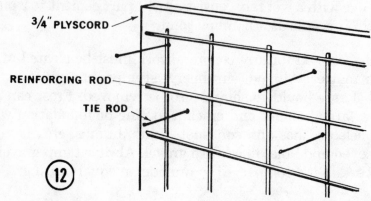

3/4" PLYSCORD

REINFORCING ROD

TIE ROD

(12)

Holes for wire spreaders are usually drilled two feet on centers. Spreaders are placed where needed to allow form to hold concrete without bulging. While a 2 x 6, stiffened every four to six feet with stakes, provides sufficient rigidity to pour a 5½ x 16" wide footing, a wider or deeper footing would require 2 x 8.

Always tamp concrete in a form. Many masons wear 16" high rubber boots, Illus. 5, and tamp concrete by walking. If you plan to make good money in your spare time doing concrete jobs for others, buy or rent a tamper, Illus. 13. This comes 36" to 48" in length. It brings fine material to the surface while it compacts the larger aggregates. Tamping can also be done by shoving a 1x4 or 2x4 into newly poured concrete to eliminate air pockets adjacent to a form or corner.

(13)

Magnesium concrete rakes, Illus. 14, are also helpful tools for amateurs who want to work like a pro on large areas. Use the teeth for spreading, the smooth edge for screeding and floating. These are available at rental stores.

Always fill a form to top edge, then screed concrete level using a 2x4 or other straight edge. Screeding any area larger than a footing form requires one man at each end of a screed. Work it back and forth. Always fill low spots with concrete.

REINFORCING WIRE AND RODS

Illus. 12, 15 show how ½" steel reinforcing rods are frequently positioned, horizontally and vertically, then wired together 1′0″ on centers. Use ¾" plyscord for forms. This is stiffened by 2x4's A, Illus. 15, 2′0″ on centers vertically. 2x4's B, spaced 2′0″ on centers horizontally, hold forms together. Steel ties C, available from your masonry supply dealer, are positioned 2′0″ on centers. These hold forms securely in place. When you strip forms, cut ties off flush with concrete.

B

A

C

When building a foundation containing a window, door or other opening, use ties at all four corners of opening. If it's a large opening, use ties every two feet as well as at corners.

When a slab requires 6x6 reinforcing wire, this refers to welded wire spaced 6″ apart vertically and horizontally. See page 32.

JOINTS — EXPANSION, CONTRACTION AND CONTROL

When laying a slab in a basement, garage, porch or patio adjacent to a house foundation, always provide an expansion joint, Illus. 16, between slab and foundation. Use ½ or ¾″ asphalt impregnated insulation board, cut to width equal to thickness of slab. Place in position shown around perimeter, also use as control joints when laying a large area; between house foundation and a sidewalk, driveway, and/or between house and garage foundation.

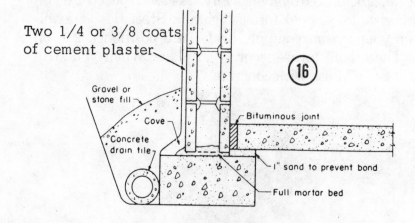

Two 1/4 or 3/8 coats of cement plaster

Gravel or stone fill

Cove

Concrete drain tile

Bituminous joint

1″ sand to prevent bond

Full mortar bed

16

Contraction joints, Illus. 17, are formed by cutting a groove 1″ deep across a slab. Allow concrete to set up sufficiently to support a 2x8 plank. Use this as a straight edge to guide a groover. This permits concrete to shrink in cold weather without cracking concrete. When laying a walk or driveway contraction joints are usually spaced every four to six feet, Illus. 18.

Two lengths of beveled clapboard, placed in position shown, Illus. 19, can also be used to form an expansion joint. Paint with motor oil. Drive 8 penny nails in thick butt edge, every 4 feet. This will permit removal after concrete begins to set. When concrete has cured, fill joint with hot tar or other bituminous joint filler, Illus. 20.

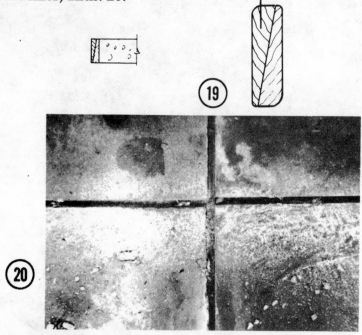

If there's any question of water seepage thru an expansion or control joint, do this. Cut strips of asphalt impregnated insulation board to width of slab, less ¾". Cut ½" strips of 1" wood and nail these temporarily to top edge, Illus. 21. 1" lumber usually measures ¾". Paint wood strips with old crankcase oil. Place expansion strips in position, Illus. 16 against forms. Keep wood strips flush with top edge of form.

WOOD STRIP

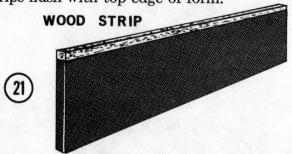

After concrete begins to set, and will support 2x8 planks without marking concrete, position planks and carefully remove wood strips. This leaves a ½x¾ joint. When concrete has been allowed to cure, allow at least a week, fill joints with bituminous sealant sold by your building materials dealer. Or you can fill joint with hot tar. When dry, sprinkle dry cement over joint, Illus. 22.

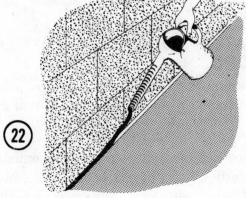

Isolate footings for columns or piers, Illus. 23, with asphalt impregnated insulation board. This prevents any movement in a slab from affecting column. When laying large areas, divide it up, Illus. 9, 24, into size sections you can handle. Use a line and line level to make certain all forms are on plane or pitch desired. Always divide a large area so you can pour and screed it easily. In this application, the asphalt impregnated strips, embedded between sections, are called Control Joints, and are always embedded permanently.

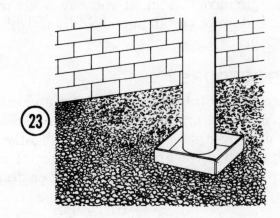

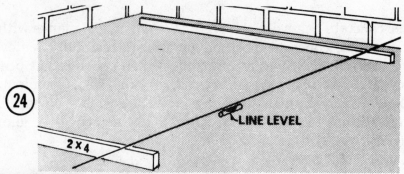

2 X 4

LINE LEVEL

ADDITIVES, WATERPROOFING, HARDENERS, DUST INHIBITORS

A concrete slab can be hardened, made comparatively non-dusting, and waterproofed with the addition of Anti Hydro Set or equal additive.

This is a good investment when pouring a basement or floor area where there's any question of dampness. It's also recommended in garages and commercial work where a thoroughly hardened and non-dusting floor is desirable.

Follow manufacturer's directions and add the exact amount of additive they specify to the water when mixing. In the case of ready mix, always pour in the exact gallons of additive required according to the size of the load. Besides waterproofing concrete, they also act as hardeners and as dust inhibitors. Measure and mix additive in exact amount of water manufacturer specifies. Then use only as much of this water to a bag of cement as specified. Cut the top off a 50 gallon drum and use this as a reservoir for your premixed water. If you buy ready-mix, add additive to tank and make certan it's mixed thoroughly before driver starts to pour.

GENERAL PROCEDURE

Pouring concrete follows this procedure. After excavating, remove all foreign matter, stumps, roots, etc. Always build forms on compacted soil or over gravel. Spray the site lightly. This is particularly helpful when you are bonding to existing concrete or stone. When pouring against existing concrete or stone paint concrete with a slush coating as described on page 17.

After filling a form, always tamp concrete to compact it. Some masons wear rubber boots or overshoes and walk around the poured concrete to make certain it is compacted particularly along the edges. The more you compact concrete, the harder it will be. Concrete in small forms can be compacted with a 2x4.

Tool rental stores stock roller tampers, Illus. 13. When doing a large floor that will be subjected to heavy loads or traffic, compacting is extremely important.

When you have filled and compacted a form, use a straight edged 2x4 as a screed, Illus. 25. Work it back and forth, saw fashion, to level concrete flush with edge of form. Magnesium concrete rakes with extension handles, Illus. 14, help place concrete in hard to reach areas. Use tooth side for screeding, smooth edge for floating.

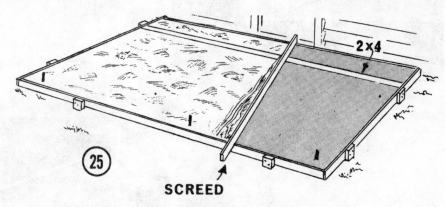

2×4

(25)

SCREED

Floating, Illus. 26, can be done with a wood float for a course texture, or with a steel float for smoother surface. The wood float produces a gritty, non-slip surface, while a steel float produces a mirror smooth finish, Illus. 27. A rougher surface can be produced by using a hair broom, Illus. 28; while a coarse brush broom, Illus. 29, produces an even rougher surface.

When working on a large area, rent a long handled magnesium bull float, Illus. 30, from your tool rental store. They also provide various finishing brooms with extension handles.

As soon as the concrete begins to set, frequently within an hour in hot weather, cut outside edge free from form using a steel float or trowel, Illus. 31. Finish edge with edger, Illus. 32. This insures a smooth compacted, rounded edge when form is removed. A long handled edger can be rented.

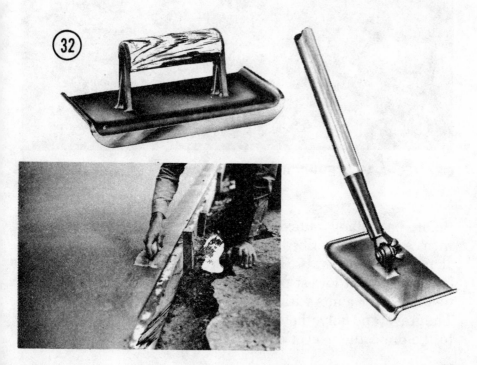

RAISE A LOW SITE

Buy fill to raise a low site. A building site must be raised to allow a natural drainoff away from building. After compacting dirt fill, cover with 6″ of 1½ gravel. This must also be compacted before setting up forms for footings.

If there's any question of ground water entering a basement or foundation, always spread sufficient 1½ gravel, plus ¾ gravel, to raise floor area above drain tile placed on outside of foundation, Illus. 16. After footing and foundation walls are completed, spread gravel to thickness job requires, then cover floor area with polyethelyne, Illus. 33. Buy sufficient width so you can lay polyethelyne in one piece. Allow it to come up sides of footings before placing expansion joints in position. Use great care not to tear polyethelyne. Spread reinforcing wire where same is recommended. Set up forms and pour floor.

GRAVEL ⤴ WATERPROOF MEMBRANE ⤴ 6″ X 6″ WIRE

Before backfilling, outside walls should be parged (plastered) with cement as described on page 36. 4″ drain tile should be laid along footing, Illus. 16. This should slope to a dry well dug as far away and at lowest point property permits. Place tile end-to-end. Cover joints with 4″ wide strips of No. 15 felt, Illus. 34. The gravel and polyetheylne should channel underground water to the drain tile. Never connect tiles to a septic tank or field.

(34)

Note page 34 concerning basements that may have a problem with water.

FOOTINGS, FLOORS, DRIVEWAYS, PATIOS

Before pouring any concrete, scrape the area clean of debris, greenery, stumps, etc. If slab is in an exposed area, i.e., patio, carport, unheated garage or driveway, where frost can get under, remove large rocks. Frost will frequently heave a concrete slab, particularly when it can use the leverage of a large stone.

Always slope a patio or porch floor, or walk, away from the house. A 3′ wide walk, sloped ⅛″ to 1′, will be ⅜″ lower on low side. When pouring a patio, a slope of ⅛″ to one foot away from house provides good drainage.

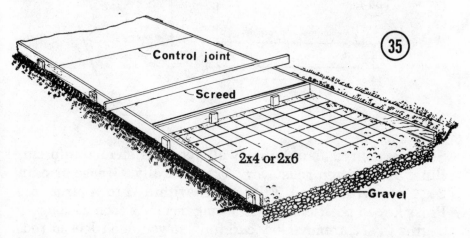

Control joint

Screed

(35)

2x4 or 2x6

Gravel

Grade site to pitch required. A driveway can be as narrow as 8' wide. A 10' width is preferable. It should pitch sideways ⅛" to a foot away from house, ⅛" per foot in overall length. A 2x4 form, Illus. 35, placed over 4" of gravel permits pouring 3½" of concrete wearing surface for a private driveway. Embed 6x6 reinforcing wire about 1" above gravel as described on page 32.

Embed ½" asphalt impregnated control strips every eight, ten or twelve feet. Divide the overall length into equal units with control strips. Note details on page 20.

Forms for a private driveway can be 2x4. If you lay a driveway serving heavy trucks, use 2x6 and pour a 5½" slab. Order ready-mix. Explain size and depth of concrete and the company will estimate quantity. Have all forms and extra manpower available when readymix truck arrives.

POURED FOUNDATION

Illus. 36 shows a keyed footing for a poured concrete foundation wall. To lock footing to foundation, make a key from 2x3 or 2x4, to length footing requires. Bevel edges to shape shown. Drill 1" holes every 12", then saw in half.

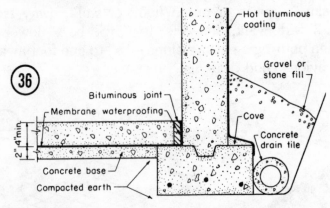

Since a poured foundation wall requires steel reinforcing, Illus. 37, position rods every 12" vertically. Grease or paint 2 x 4 key with old oil. Fill form to within ½ to ¾" from top. Press key in position. Finish filling form. As soon as concrete begins to set, remove key carefully so you don't loosen rods.

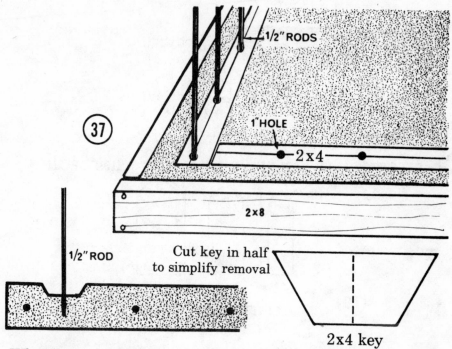

1/2" RODS

1" HOLE

2 x 4

37

2 x 8

1/2" ROD

Cut key in half
to simplify removal

2x4 key

When footing has been allowed to set 3 days, wire ½" reinforcing rods horizontally every 12" to vertical rods, Illus. 12.

FLOORS OVER WET AREAS

Illus. 36, 38, show sub-slab recommended where dampness may cause a problem. Excavate 2" below top edge of footing. Paint edge of footing with a bonding mixture containing one part cement to two parts sand. Lay 1" of concrete. Spread 6 x 6 wire, Illus. 39. Lay another 1" of concrete.

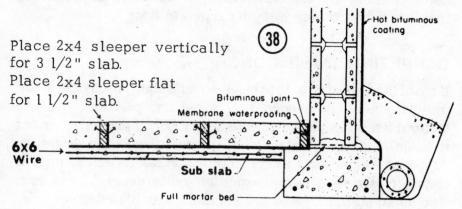

Place 2x4 sleeper vertically
for 3 1/2" slab.
Place 2x4 sleeper flat
for 1 1/2" slab.

38

Hot bituminous coating

Bituminous joint

Membrane waterproofing

6x6 Wire

Sub slab

Full mortar bed

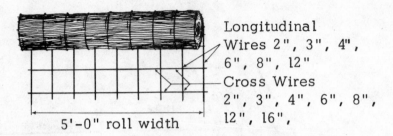

Longitudinal
Wires 2", 3", 4",
6", 8", 12"
Cross Wires
2", 3", 4", 6", 8",
12", 16",

5'-0" roll width

ELECTRICALLY WELDED WIRE FABRIC REINFORCING

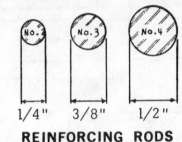

1/4"   3/8"   1/2"

(39)   **REINFORCING RODS**

Screed 2" sub-slab flush with form. Allow this slab to cure thoroughly then mop it with hot tar. Lay and carefully roll No. 15 felt into the wet tar so felt bonds to slab. Rent a roller. Paint 2" of each strip and overlap each course 2". Allow felt to come up sides of foundation about 6". When you have covered the slab completely east to west, apply a coat of hot tar and lay felt north to south. Again overlap each course. Mop surface with hot tar. Allow to dry. Next place expansion joint in position using either the beveled siding or asphalt impregnated insulation board before laying concrete floor.

CARPETING OVER CONCRETE

If a finished wood floor is to be laid over a slab, or particle board underlayment is needed for carpeting, position pressure treated 2 x 4 sleepers, Illus. 38, every 16". If pressure treated lumber isn't available, paint sleepers with wood preservative. Allow preservative to dry thoroughly before embedding in concrete. Drive 8 penny common nails every 16" to lock sleeper to slab. Place first 2 x 4 along outside edge.

Use sleepers as forms and pour concrete flush with top. Use extreme care not to puncture waterproofing membrane. A 1½" thick slab is ample for a floor over a sub-slab where there's no underground water pressure.

After slab has been poured and allowed to set, cut felt flush and seal joint with hot tar.

If you are laying a concrete floor over a hot water heating pipe or water line, don't cover pipe with concrete, Illus. 40. Build a form to keep concrete at least 6" from each side of pipe. Carefully screen sand to make certain it doesn't contain any nails, cinders, or other foreign matter. Fill area around pipe with sand. Draw a chart showing exact location of pipe. If at some future date, repairs are needed, you will not only know where to find the pipe, but will also cut repair costs to a minimum.

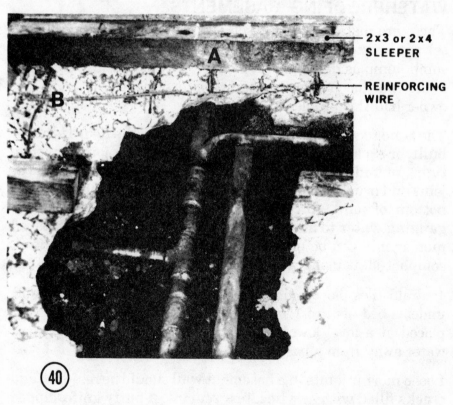

2x3 or 2x4 SLEEPER

REINFORCING WIRE

LOOK BEFORE YOU BUY

Good construction depends on what you know and do. Unlike the purchase of a new car that turns out to be a lemon, a poorly constructed house can be a health hazard if underground seepage penetrates a basement or crawl space. A poorly installed septic tank field, improper drainage from house leaders, or septic tank drainage off a neighboring property can make a house untenable.

Houses built on slabs in areas where each is serviced by a septic tank, will sometimes develop an odor no rose can negate. Use caution, particularly when purchasing a house in a lower section than others. Note whether owner uses a lot of air freshener. Few people think to inspect real estate with their nose, and yet a nose knows.

WATERPROOFING BASEMENTS

There are many ways to resolve this problem. Some builders solve it by expecting the owner to "live with it." They install a sump pump as described on page 77. If you consider buying a house that has a sump pump in the basement, you know the owner had this problem.

The time to waterproof a basement is when the house is being built, or excavate to do the same job. All mortar joints must be tight, outside, as well as inside. Remove loose mortar. Wet joint and brush on a slush coat. With wood float placed against bottom of joint, force mortar in. Use a small trowel. Use gauging water to make a plastic mix, Illus. 41, or a prepared mortar mix can be used. When mortar sets fingerprint hard, compact joint using a concave jointer, Illus. 42.

4″ drain tiles placed in position indicated, Illus. 38, over and under a bed of ¾ to 1½ gravel, and sloped towards a dry well, placed at a level lower than the footing, will normally carry water away from a basement.

Loose mortar joints in a basement wall should be removed and cracks filled with acrylic latex sealant. A putty knife, dipped in water, can be used to smooth sealant flush with surface.

TUCKPOINTING BLOCKS, CHIMNEYS, ETC.

Check your chimney for cracks caused by settling, freezing, etc. Always seal even the smallest crack with acrylic latex sealer.

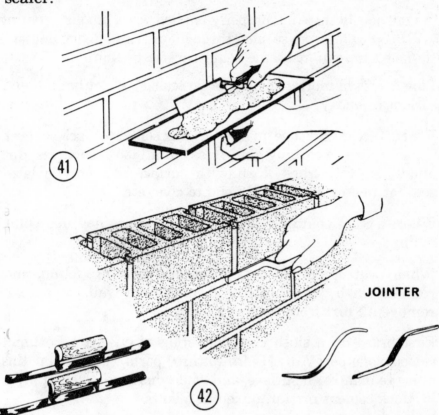

JOINTER

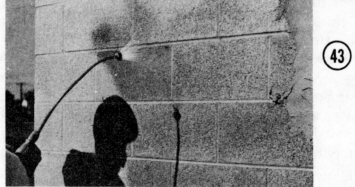

To insure bonding a parging coat to surface, wire brush surface clean, then spray with clean water, prior to waterproofing wall, Illus. 43.

If you are attempting to waterproof a basement where there's water, the water must be pumped out, and kept out, to allow the coating to set up for at least 24 hours. The pump must not carry out any slush coat or mortar.

As previously stated, a properly applied waterproofing coating will keep water from seeping through walls and floors, but isn't designed to reinforce same against water pressure.

Fine cracks in mortar joints can be sealed with either a liquid sealant or epoxy.

For surface cracks, use an acrylic latex concrete crack sealant. This requires a calking gun. Hose out all loose particles, dirt and dust. Fill crack flush with surface. An acrylic latex sealant makes a chemical bond to concrete.

Where a crack contains water that can't be drained, use a plug sealant.

When waterproofing outside of an existing foundation, remove earth, wire brush and hose down wall. Be sure to remove all dirt from mortar joints.

First brush on a slush coat consisting of one part Portland cement, one part Anti Hydro to three parts water. Paint this over the dampened surface, and while this is still wet, apply a ⅜" thick cement mortar, mixed as follows.

Mix the equivalent of one part of Anti Hydro to 10 parts of water. This is called gauging water. Mix one part Portland cement to two parts of clear sand. Add gauging water. Apply a ⅜" thick coating to walls, Illus. 44, to one foot above grade line. Do a two coat job. Scratch the first coat, Illus. 45. Use a trowel or screwdriver. Allow it to set thoroughly, spray before applying a ⅜" finishing coat, Illus. 46. Use the same mix as scratch coat. Smooth finish this coat. Allow each to cure at least a week before applying the next coat. Spray it each day with a fine mist to help cure it properly. When cured and dry, apply asphalt cement, Illus. 47. Embed No. 15 felt horizontally in the wet asphalt. Start at bottom and overlap each course 6", Illus. 48.

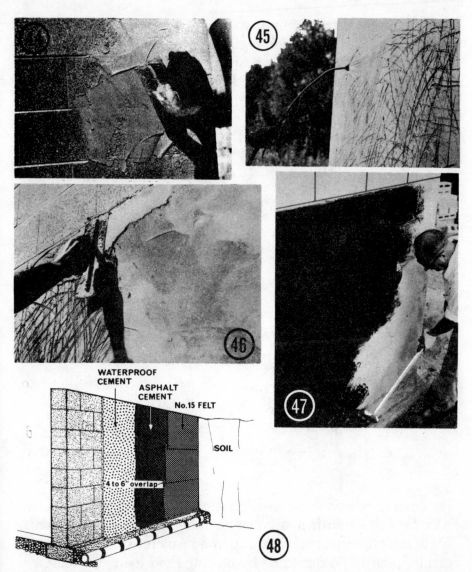

Paint with hot tar or asphalt cement to thoroughly seal all joints. After allowing to dry, back fill by hand using care not to bruise or break the membrane.

When finishing a basement floor, lay it as suggested on page 31. If you are refinishing an existing one, clean concrete floor, bond a waterproof membrane as described on page 32, then top it with minimum of 1″ concrete. Use previously recommended mix. Trowel smooth.

To prevent cold from penetrating a floor slab, insulate around perimeter. Place 1" thick rigid asphalt impregnated insulation or fiberglass board in position shown. Cut to width equal to thickness of slab edgewise, Illus. 49. Cut another panel 16" wide. Place in position shown around perimeter.

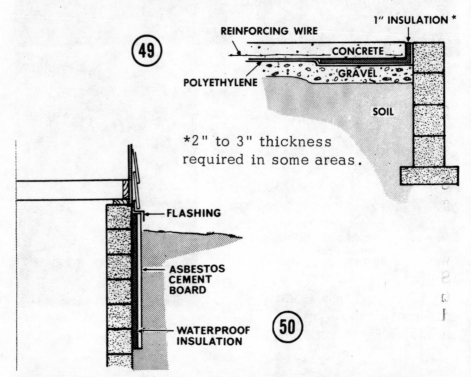

Basements where condensation creates a water problem can now be solved with a number of different types of sealants. Moisture buildup on a block or poured concrete foundation wall can frequently be remedied by placing rigid foam, or fiberglass insulation against the outside wall, Illus. 50. Use 1", 2", or thickness required to protect against temperature in area. This insulation should be bonded to wall with asphalt cement and should cover wall down below frost level. Cover insulation with ¼" asbestos cement board. Cut a strip of aluminum or copper flashing. Insert flashing under siding, or under siding starter strip, then bend it over asbestos cement panel board. This keeps rain from penetrating insulation. Backfill and your basement should be degrees warmer.

Water seepage into a basement can usually be traced to concealed, or exposed cracks in mortar joints, in cracked blocks, or from pressure building up alongside, or under basement floor. Two of the most common causes are:

1. A poorly installed leader carrying water from gutters. If leader is clogged, a joint loose, or leader discharges too close to foundation wall, it creates problems that can easily be rectified.

2. Hard to find hairline cracks in mortar joints can frequently be located when you apply sufficient water at grade level.

Check each leader to make certain it drains away from foundation to either a run-off, or to a dry well. If you build a dry well, do so far enough from foundation and at a level that can absorb it.

Seepage through cracks in mortar joints, can be sealed in several different ways. If crack starts near grade level, Illus. 51, dig a trench with bottom of trench level with crack. Cover side, but not bottom of trench with polyethylene.* Soak trench with water so it filters into, and soaks crack. Next pour in Hydro Stop, or equal ready to use liquid sealant. Pour sealant into trench and it will seep into and fill crack. Since there are many different types of liquid sealants available, follow manufacturer's directions.

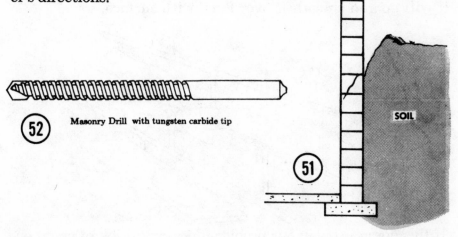

(52) Masonry Drill with tungsten carbide tip

SOIL

(51)

*Polyethylene wrappers from your dry cleaner can be used.

39

Keep pouring sealant into trench until foundation refuses to absorb it. Wait awhile, then apply a fine mist of water and flush any sealant remaining in trench. Some of this will be absorbed by foundation.

Some hairline cracks frequently require a second application. Apply same 24 hours later. Again, wet trench with water to soak crack before applying second application.

If a crack starts below grade, where it's concealed on the outside by a concrete walk, but shows up inside basement, you can waterproof it in two ways. Using a carbide tipped masonry bit, Illus. 52, rout out mortar at least ½" to ¾" deep where crack appears. Remove all loose particles. If you don't have an electric drill, borrow one, or use a can opener. Make a V groove full length of crack and at least ½" to ¾" deep.

Apply a paste sealant, either latex or epoxy, Illus. 53. Some latex based patching material comes in two parts, a liquid latex and a dry powder. Mix only as much as you can apply immediately after mixing. This paste sets up fast — some within two minutes, so follow manufacturer's directions and you'll waterproof like a "pro." Apply patching paste with a putty knife. Working as quickly as possible, work it in as far as it will go and smooth it over flush with surface.

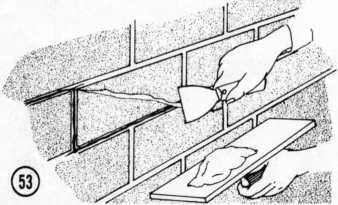

If this doesn't solve your problem, locate position of crack on inside by measuring over to a window or door and from top of foundation down. Do same outside. Drill a ½" or larger hole through concrete, Illus. 54, then drive a rod down to a depth just above crack. Remove rod and carefully insert a piece of copper tubing. Fasten a piece of rubber hose and funnel to top end, Illus. 55. Fill funnel with water. When crack is saturated, pour in liquid sealant.

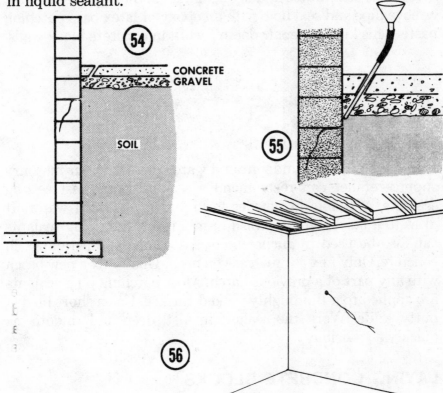

If you have a long crack, or if a portion of the foundation wall is allowing water to enter at several different levels, draw a chart to accurately locate area, then drive a crow bar or iron rod to depth required to service each course.

Cracks that take in a lot of water should be sealed inside with paste sealant, and on outside with liquid sealer.

Another way of filling a crack below grade in a poured concrete wall, one that shows up on inside of basement, Illus. 56, is by drilling a ½″ or larger hole at top of crack. Drill hole at a slight downward angle to a distance halfway through wall. Use a ½″ or larger masonry bit.

Insert a piece of ½″ copper tubing of sufficient length to penetrate halfway into wall and still project at least 2″ from wall. Use rubber tube and funnel, Illus. 57. Apply water, then sealant as described above. Keep pouring until it won't take any more. Wait 15 to 20 minutes and try again. If it still won't take more, remove pipe and seal hole with an epoxy or latex base patching paste. This kind of repair doesn't work in concrete block walls.

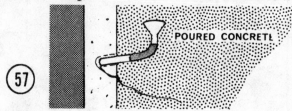

POURED CONCRETE

(57)

Even where there's underground water pressure your masonry supply retailer can recommend a sealant. Some fast setting sealants, available in powder form, are mixed with water. It dries to a hard metallic finish in minutes. These plug sealants can also be used to anchor bolts, fasteners or iron railings in concrete. Only mix as much as you need. Don't mix a new batch with any part of a previous batch. Most patching plug sealants are applied to a thoroughly soaked surface. Use a short bladed putty knife. Work the sealant in fast, deep and smooth, as quickly as possible.

LAYING CONCRETE BLOCKS

Laying blocks for a wall requires:
1. Selecting a site that's within your property line, one that doesn't prevent a natural run-off of water; or diverts water into your neighbor's property.

2. Clearing area and excavating to a depth equal to, or below frost level.

3. Erecting forms and pouring footings, Illus. 58.

If you plan on running water, electric, gas or telephone lines through foundation, make an open end box from four pieces of 2x6 or 2x8 x 16″ if lines are going through footings, 8″ long if lines go through foundation wall. Or you can use 4" diameter drain tile, Illus. 58. Place in position indicated. Be sure to mark location and depth of this opening on your foundation plan. If you use lumber for a form, knock out the form after concrete has set.

Always purchase extra blocks to allow for breakage. When necessary, make heavier mortar joints between blocks to fill space available. When estimating the number of blocks required, figure three blocks per course to every 4'.

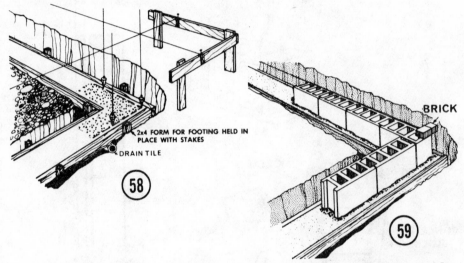

2x4 FORM FOR FOOTING HELD IN PLACE WITH STAKES

DRAIN TILE

BRICK

58

59

4. Locate exact corners. Drop a plumb bob down from guide lines.

5. Lay out a course of block dry, no mortar, to establish exact number of blocks needed for first course. Use a ⅜" piece of plywood for a ⅜" thick mortar joint. Use a ½" piece for ½" joints. ⅜" is the preferred size. Cut a block if necessary.

Use end blocks on exposed ends. Always stagger joints on adjacent courses. See page 124, for complete chart showing various size and shape blocks.

6. After establishing the number of blocks for first course, and exact size of mortar joints, remove blocks, spread an inch layer of mortar on footings, Illus. 58, at corners of one wall.

7. Drop a plumb bob down from guide line and allow point of bob to mark concrete at three points. Draw lines in mortar to indicate corner. Place corner block. Check block with level horizontally and vertically. Do the same at the opposite corner.

Stretch a line between corner blocks, Illus. 59.

If you need to cut a block to fill a course, use a chisel, Illus. 60. Draw a line on block, Illus. 170. Strike chisel with hammer.

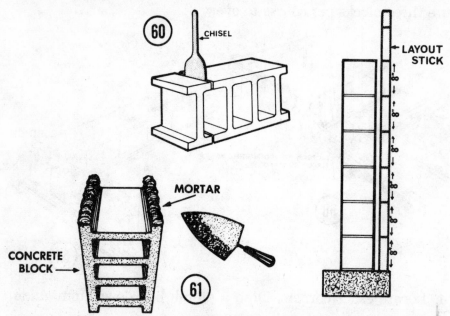

After spreading mortar on footing, butter up end of each block, Illus. 61, and shove it into position. Keep all mortar joints to thickness estimated. If necessary to set some blocks with a thicker mortar joint to fill space, a ½″ or ¾″ mortar joint isn't unreasonable, providing you stagger all joints as shown, Illus. 62.

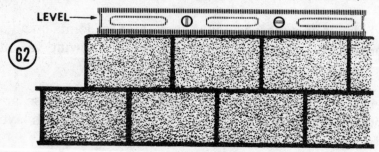

Tap each block into position. Never move a block after concrete starts to set.

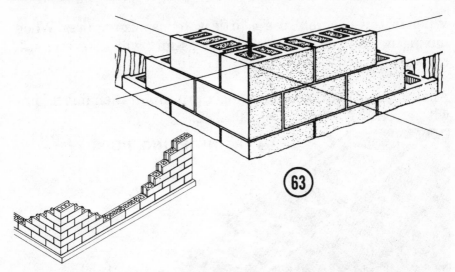

Build up corners, Illus. 63, to height required. Always keep guide line taut. Using a trowel, cut surplus mortar away from joint and throw it back onto mortar board, Illus. 64. Always turn mortar with your trowel so no small globs harden.

Use a ⅝ or ¾ piece of plywood as a mortar board. Wet board before placing mortar, or use a wheelbarrow.

Never mix any more mortar than you can conveniently use within 2½ hours. On a hot day, add water to keep mortar plastic.

When mixing concrete, use as little water as mix requires. When mixing mortar, use as much water as mix will take and still remain plastic.

Allow mortar joints to set up finger print hard then finish joint with a concave or V jointer, Illus. 42.

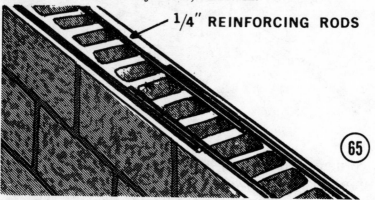

1/4" REINFORCING RODS

(65)

Use reinforcing rods, Illus. 65, or welded wire, Illus. 66, when building a wall where ground pressure requires same. When laying rods, overlap ends.

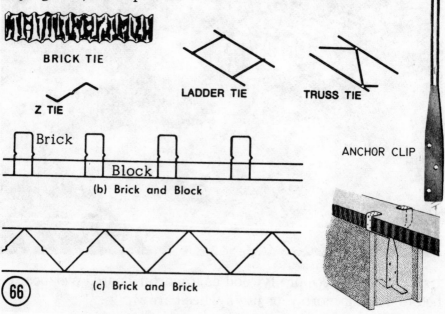

BRICK TIE

Z TIE

LADDER TIE

TRUSS TIE

ANCHOR CLIP

Brick

Block

(b) Brick and Block

(c) Brick and Brick

(66)

When building a concrete block wall that is to be faced with brick, use 10"x8"x16" blocks below grade, Illus. 67, then 6"x 8"x16" blocks from grade level up. This allows a shoulder for brick facing.

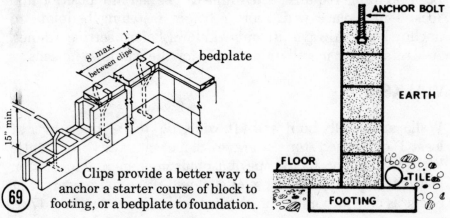

(67)

6"→

10"→

Brick Veneer Tie Spacing

Design Wind Load, psf	Spacing Horizontal by Vertical, in.	Wall Area per Tie, sq ft
20	24 by 24	4
30	16 by 24	2 2/3
40	16 by 18	2

(68)

When you plan on brick facing a concrete block wall, embed brick ties, Illus. 66, in position indicated, Illus. 68.

When building a foundation, embed anchor bolts in top course. Most masons set anchor bolts 4" in from outside edge of foundation wall, in position indicated, Illus. 69. A washer at head of a bolt locks bolt securely in concrete. Fill entire core with concrete allowing only threaded end to project about 4" above block. To hold bolt in place, drill a hole in 2 x 4. Lay same across block. Snap a chalk line and embed bolts to line.

bedplate

8' max.
between clips
1'

15" min.

ANCHOR BOLT

EARTH

FLOOR

TILE

(69) Clips provide a better way to anchor a starter course of block to footing, or a bedplate to foundation.

FOOTING

If site selected contained any soft spots that needed to be removed, and filled with gravel, place ½" reinforcing rods the entire length of footing, in position shown, Illus. 70. This beefs up footing. Overlap ends of rods at least 4".

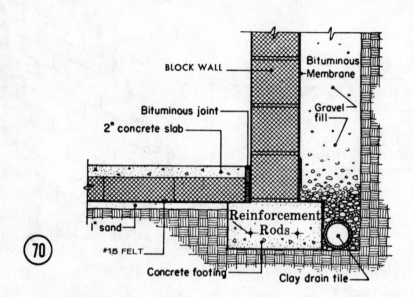

BLOCK WALL

Bituminous Membrane

Bituminous joint

2" concrete slab

Gravel fill

Reinforcement Rods

1" sand

70

#15 FELT

Concrete footing

Clay drain tile

Never lay concrete footings or block in freezing weather unless anti-freeze is used in the mortar and the work is covered at night to protect against freezing or snow.

If you are building a basement, and plan on installing a steel or aluminum window frame, use slotted blocks, page 103 , in position frame requires. The frame can be slid into position and locked into place with mortar before fastening bedplate to anchor bolts. Always lift out windows before setting frames. Ask retailer for installation details for the windows he sells.

WALKS

Walks are usually built 3 to 4 ft. wide, 1½" to 3½" thick, over at least 3" of crushed stone or gravel. Slope walk away from house ⅛" to ¼" per foot. Cut strips of asphalt impregnated insulation board 1" less than thickness of slab and place these every 4 or 6 feet as control joints. Finish joint with a groover, Illus. 17.

Forms for walks can be built as shown, Illus. 71. Pour concrete in alternate sections. After these sections set up, remove end forms (don't disturb expansion joint) and pour intermediate sections. Screed concrete flush with top of form using a straight 2x4 as a screed. Work this back and forth, saw fashion. Use a smaller screed for smaller area.

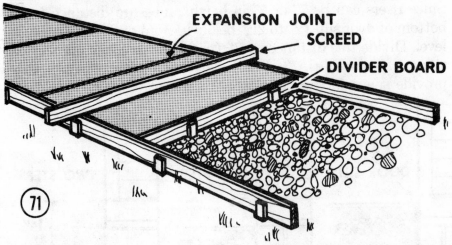

EXPANSION JOINT
SCREED
DIVIDER BOARD

A self-aligning screed that keeps working concrete to center of form, instead of over edges, may be built as shown, Illus. 72. Bolt 2 — 1x4 by length needed. Nail 3″ blocks in position indicated. Spread concrete by working this screed back and forth.

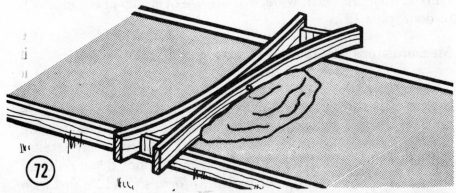

Allow concrete to set until stiff enough to float. Follow directions previously outlined. Use a steel or wood float to obtain finish desired. Don't overwork the surface as this tends to make a chalky, dusty surface. For a gritty, nonskid surface, use a broom, page 26, to roughen surface.

BUILD STEPS

Build forms for steps with lumber free of knots and smooth on one side. Keep smooth side in. Brace forms with stakes to prevent buckling. Excavate area for steps to depth below frost level. Fill with stone and gravel.

Since risers can be 7" or 8" in height, measure distance from bottom of door sill, (2" to 2½" below bottom of door), to grade level. Divide this distance by 7" (or 8") to estimate number of steps or risers, Illus. 73. Build platform to height required to provide space for needed number of steps. Check with level.

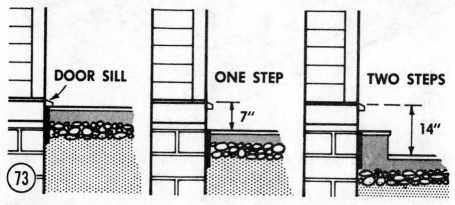

Steps usually extend beyond edge of door trim, same width of trim. If door trim is 6" wide, overall size of step will equal width of door, plus 12".

Measure steps you find acceptable, build yours same size.

The starting platform should measure an additional 12" wide. Step treads can be 8" wide when overall distance X, Illus 74, is 6'0" or less; 10" wide when X measures up to 8'0". The starting platform should be constructed to height required. Drive stakes firmly into ground. Nail stakes to B but do not drive nails all the way in. The top edge of platform form should be distance down from bottom of door sill to allow for number of steps required. Use 1x8 for a 7¼" riser. Use 2x8x8 or longer for forms. Use 1x4 for cross ties to prevent forms from buckling. To permit finishing steps under riser, Illus. 75, bevel bottom edge of A to angle shown.

Build form for platform, Illus. 74, to height above grade needed for one, two, or more steps. Excavate to depth needed if you want surface of platform flush with grade, or to any height above.

Drive stakes in ground to hold forms in position. If stakes are driven in flush with top edge of form they don't interfere when you screed the concrete.

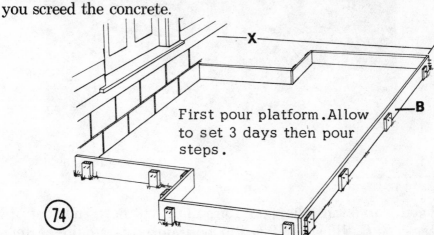

First pour platform. Allow to set 3 days then pour steps.

Build form for steps using 1x8 or ⅝″ plyscord. Cut to overall size required , Illus. 75. Measure 5½″, 6″ or 7¼″ down for one step, 11″, 12″ or 14½″ down for two steps, Illus. 75. Pour concrete for steps in one pouring. Work concrete down with beveled 1x4 or pointed 2x4. Use mix specified, Illus. 1.

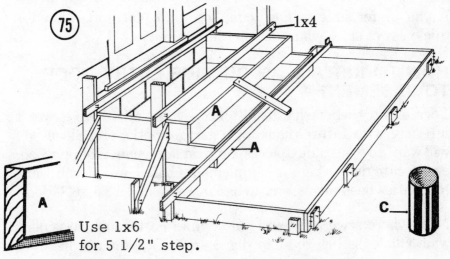

1x4

Use 1x6 for 5 1/2″ step.

See Page 59.

(76)

If you plan on installing an iron railing, cut both ends out of beer cans C, Illus. 75. Place in position posts require. After concrete has been allowed to set, remove cans by bending with pliers. When you get ready to fasten the railing in place, Illus. 76, fasten it to the house and brace the posts in plumb position. Check with level vertically as well as horizontally. Secure posts in position with lead or epoxy cement.

If you prefer slate or flagstone steps and platform, allow for thickness of slate when figuring height of riser.

HOW TO INSTALL AN OUTSIDE ENTRY DOOR TO BASEMENT

Locate entry where it's readily accessible and where it won't interfere with future plans for a garage, patio, etc. Since you will want to store bulky pieces, carry in large panels of plywood, etc., locate door where it complements basement. If you want to store a wheelbarrow, etc., in basement, install a 3'0"x6'8" door.

Note where sewer, gas and water lines go through basement wall. Don't disturb these or dig near a septic tank or field.

If you have a basement window, consider removing same and cutting door through. Don't install a door closer than 2'10" from a basement window. After selecting location, ask your lumber dealer for size of pre-finished basement entry door he has available. He will advise size of excavation, foundation, and size of rough opening. Excavate to depth equal to bottom of house footing.

Working inside basement, draw a line to indicate center of proposed opening. Draw lines equal distance from center line, Illus. 77 to indicate width of opening. Go outside and stake out site of excavation, Illus. 78.

Cut opening width of door plus 5½". This allows for 2 x 6 or 2 x 8 frame.

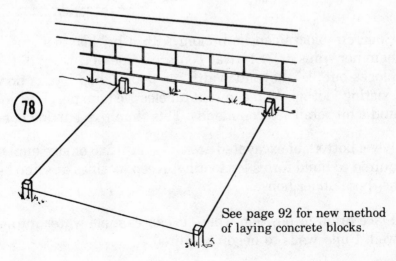

See page 92 for new method of laying concrete blocks.

Excavate to depth of house footing, Illus. 79. If a prefab stair stringer is to be installed, excavate to size stair stringer manufacturer specifies. If you build your own stairs, slope hole as shown, Illus. 80.

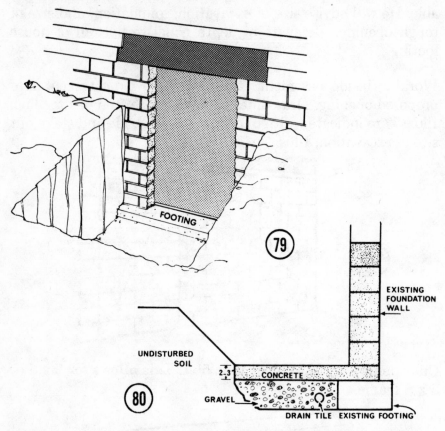

Concrete blocks can be broken with a bricklayer's chisel and hammer, Illus. 60. Start at center of line and work out. Chip blocks out 5½″ beyond width of door. Chop blocks down to existing footing. You can rent an electric hammer. Use goggles and a bit retailer recommends. This simplifies cutting opening.

Level bottom of excavated area. Use 2x4, 2x6 or size lumber required to build forms for footing. Keep footings at same height as those under house.

Place drain in position where it can channel water into a dry well. Build walls to height required.

If you prefer concrete block steps, Illus. 81, use 8x10x16" or 18" blocks.

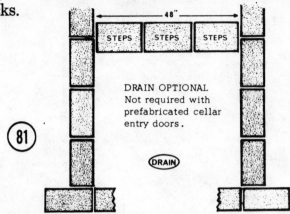

48"

STEPS STEPS STEPS

DRAIN OPTIONAL
Not required with
prefabricated cellar
entry doors.

(81)

(DRAIN)

Back wall of excavation can be sloped as shown, Illus. 82. Cover undisturbed soil with 2" of gravel, and set blocks in a thick bed of mortar.

8" CONCRETE BLOCK SIDE WALL

STEP

STEP

STEP

CONCRETE

STEP

STEP

(82)

CONCRETE

CELLAR FLOOR

FOOTING

DRAIN

Allow each step to overlap step below 1", Illus. 83.

(83)

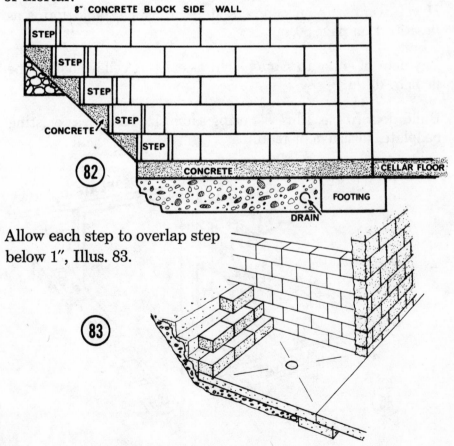

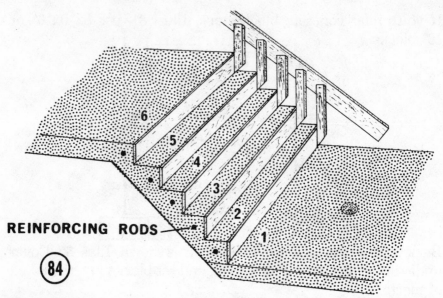

REINFORCING RODS

84

If you want to pour concrete steps, Illus. 84, build forms as described on page 50.

Lay floor of entry to same height as existing cellar floor. Slope floor to drain.

Build door frame, Illus, 85, using same size lumber as existing bedplate, to size door requires.

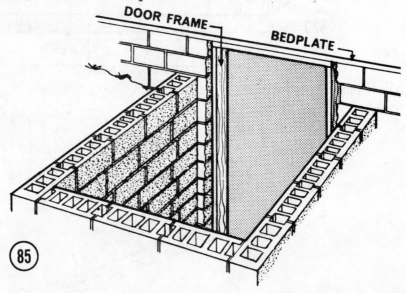

DOOR FRAME

BEDPLATE

85

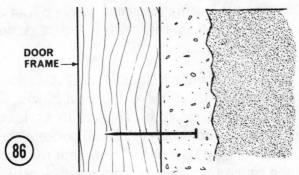

DOOR FRAME →

(86)

Drive 8 penny common nails into frame, Illus. 86, to anchor frame in concrete.

Set frame in place, flush with inside face of wall. Check with level and brace in position, Illus. 87.

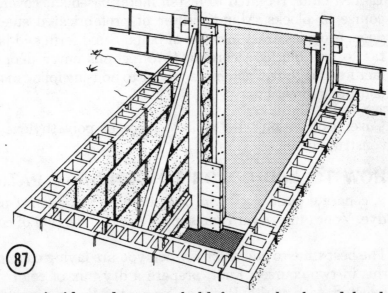

(87)

Nail braces, inside and out, to hold frame plumb and level against block wall. Nail through header to bedplate. If header doesn't butt against bed plate, cut 2x4 filler blocks to length required and nail to header and bedplate.

If two floor joists are nailed together within area cut for door, use a double header over door. Use two 2x4 studs between header and bedplate. Place 2x4 studs under bedplate, in line with floor joists.

Wet edge of block and footing adjacent to door frame. Paint broken edge of blocks and footing with a 1 part cement to 2 parts sand wet cement mix.

Temporarily nail 1, 2 or 3 short lengths of 1x6 to inside and outside of door frame. Brace other end against blocks, Illus. 87. Fill form with 1:2¼:3 concrete mix. Tap 1x6 to settle concrete. Use a stick to make certain concrete fills all pockets. Add 1x6's as you work your way up. Use a stiff concrete mix when you get to top. When concrete begins to harden sufficiently, remove forms and plaster surface to desired smoothness.

Install door using three 3½"x3½" hinges. Nail inside casings around door. Nail door stops to frame in position required.

Embed bolts, required to fasten metal basement cover, in top course of blocks. Manufacturer of prefabricated steel entry doors provides exact location of bolts. Build form and lay mortar to cap blocks according to directions entry door manufacturer suggests. Fasten entry door to housing following manufacturer's directions.

Cover excavation with a large piece of polyethylene during construction.

HOW TO ADD COLOR TO CONCRETE PATIO

A concrete patio deck can be painted with masonry paint or dye. Your lumber yard can supply a wide selection of colors.

The best time to add color is when you are laying concrete. Do this in two courses. First, prepare a dry mix of color powder selected with white Portland cement. Mix in proportion to manufacturer's recommendations. Do not add water; keep dry.

Lay first course of concrete to within ½" to 1" from top of form. Add dry sand to your color mix in proportion recommended, mix well. When thoroughly blended, add water. Spread and level color course following procedure outlined. Use color powders full strength or less in proportion per bag of cement to obtain pastel shades.

ORNAMENTAL IRONWORK

Handsome ornamental railings, porch and patio posts, are now available at your lumber dealer. These should be set in place before pouring concrete. Locate exact position of posts and place a beer or coke can in position to receive base of ironwork. Cut both ends off so you have a smooth cylinder. Grease outside of can. Place cans in position posts on railing require. After pouring concrete, remove can as soon as concrete permits.

After concrete has been allowed to set (three or more days), position ironwork and fasten in position following manufacturer's directions. Either set ironwork in melted lead, or use the plug sealant mentioned on page 40.

If a leg on iron railing has rusted out, hacksaw leg close to railing. Leave sufficient stud above concrete to grasp with pliers. Use a propane torch, Illus. 88, to heat lead. Use pliers to remove embedded piece. Plug cement sealants or hot lead can be used to anchor new support.

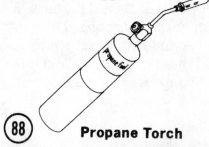

(88) **Propane Torch**

In many cases a piece of 3/16 x 1" strap steel, bent to shape shown, Illus. 89, drilled with holes indicated, can be embedded in lead or plug sealant. Drill holes in lower rail to match those in new leg. Bolt leg to rail.

Where an existing leg can't easily be removed, drill through bottom rail with a 5/16, ⅜ or ½" bit. After drilling hole through rail, use a carbon tipped bit to drill a hole in concrete. Insert a threaded steel rod through rail, thread a nut on rod. Drive rod into hole. Fasten rod in hole with lead. When lead sets, snug up nut to bottom of rail. Place another nut, Illus. 89, in position shown and cut surplus rod with a hacksaw.

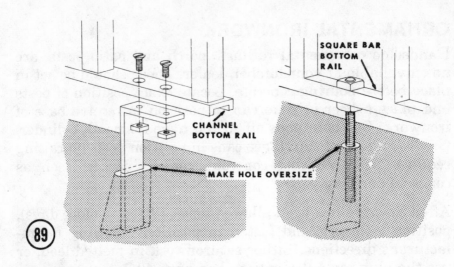

SQUARE BAR BOTTOM RAIL

CHANNEL BOTTOM RAIL

MAKE HOLE OVERSIZE

89

FRONT STEP REPAIRS

Repairing cracked and broken concrete steps is no big deal. It does take a bit of time to make a form to fit the curve of your step.

To shape a rounded edge, cut a piece of ⅛" hardboard to exact shape. First cut a paper or cardboard pattern. When you have it, trace to hardboard. Fasten hardboard to a 1 x 2 form, Illus. 90. Cut 1 x 4 or 1 x 6 A and B to exact width required. Use a concrete block to hold these in position. Mix mortar so it holds its shape. Apply and shape mortar with form.

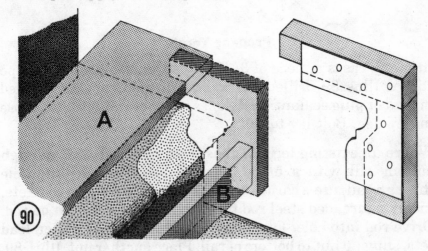

A

B

90

PAVE A PATIO

Illus. 91 shows various designs that can be achieved by either grooving joints, or using ¼ x 1" aluminum divider strips. A similar effect can be achieved by embedding 1 x 4 redwood strips in concrete. To simplify explaining procedure, a patio measuring 9 x 12 is illustrated. You can build to any other size by following the same procedure.

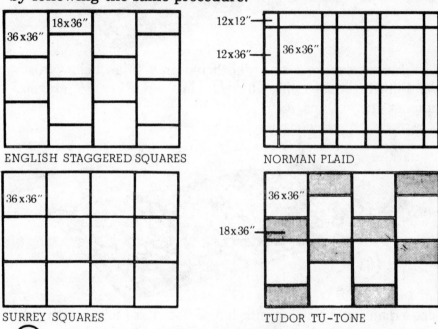

ENGLISH STAGGERED SQUARES

NORMAN PLAID

SURREY SQUARES

TUDOR TU-TONE

(91) **PATIO DESIGNS**

Place forms at a height that slopes ⅛" per foot away from house and finishes at least 1" above grade at the lowest point. A 9'0" wide patio will be 1⅛" lower on outside edge.

Drive stakes to indicate area to be excavated. Tie guide lines temporarily to stakes. After driving stakes into four corners, measure diagonals. The lines will be square when diagonals are equal in length. Drive a nail into each corner stake to indicate exact corner.

The deck should be laid in two pourings. Each slab can be 1½" thick. Use 2x4 flatwise. To allow for forms excavate area 3½" wider and 7" longer than actual patio.

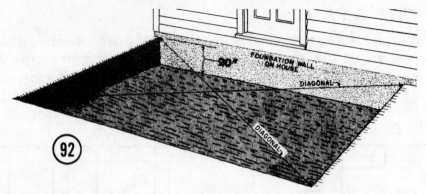

(92)

In cold areas excavate to a depth required to go below frost level. Fill with small stone, Illus. 92. Level with ¾ or 1½" gravel. Compact gravel, Illus. 93.

STONE

(93)

Place expansion strip alongside house. Paint 2x4 forms with old crankcase oil and place in position. Keep inside dimension of form 9′0″x12′0″ or size desired. Use stakes to hold forms in position. Be sure to slope form to pitch desired. Backfill form to keep it from losing concrete.

Cut reinforcing wire approximately 3″ less in overall size than form. This permits keeping wire away from aluminum. Raise wire 1″ with small stones or globs of concrete, Illus. 94. Pour sub-slab using 1 part cement, 2¼ parts sand, to 3 parts gravel. Screed concrete level with form using a straight length of 2x4.

When concrete starts to set, place a 2x8 plank across slab and remove 2x4 form adjacent to house. Don't remove expansion joint. Fill void with concrete. Score concrete using a sharp stick or trowel. Spray slab daily to cure it properly.

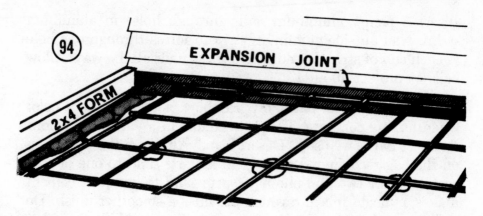

Next lay out aluminum divider strips in the design you select, Illus. 95.

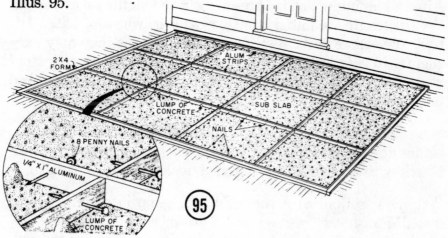

The strip adjacent to house foundation butts against ½" expansion joint. Nail aluminum strip in position flush with edge of expansion joint, Illus. 94, 95.

Aluminum strips are available in 6' and 8' lengths. These not only simplify laying and screeding a patio, but also add a "terrazzo effect." Figure the number of aluminum strips required for size and design of patio you want. Bore holes to receive 8 penny aluminum nails. Space holes about 8 inches apart.

Lay a second 2x4 form flatwise on top of first form. Position aluminum divider strips in pattern desired. Use globs of concrete to hold divider strips in position and level with form.

Insert 8 penny aluminum nails through holes in aluminum strips. Nail outside aluminum strip to forms through one hole at each end of strip. Do not drive these nails all the way. Allow nails to project about 1″.

To create a colorful patio, mix rainbow or equal cement color when mixing concrete. This is available in a variety of colors. In a test patio, we used "Desert Tan." Although the directions on the bag recommend adding three parts sand to one of concrete; two or two and one-half parts sand to one part cement makes a "neat" mixture which provides a smoother finish. Do NOT use salt beach sand as the salt in the sand kills the color.

After all aluminum strips are in position lengthwise across patio, cut lengths required to fit across 9 ft. width of patio. It's best to measure and cut strips one at a time as they may vary slightly in length. Place strips on edge, correctly spaced. Hold strips in correct position with lumps of cement as described previously.

When positioning aluminum, make certain each is straight, slopes desired pitch, and is even with top edge of 2x4 form. Check with a taut string stretched across top edge of form.

Pour two sections at a time. Spread, tamp into corners, and screed. Finish each section level with edge of aluminum strip. Work a 1x4x4 ft. straight edge across aluminum strips to level, Illus. 96. Float rough with a wood float. Float smooth with a steel float. If the concrete is on the dry side, you can begin to float each section almost immediately.

(96)

Cover with polyethylene, #15 felt or building paper, to protect patio from sun while concrete is setting. Spray patio once or twice a day to help cure properly.

After concrete has set thoroughly, remove 2x4 forms. Polish top edge of aluminum strips with a fine steel wool. Back fill around exposed edges.

To give patio a glossy and water-shedding finish, apply a coat of liquid floor wax. Polish with an electric waxer. Repeat polishing procedure with a second coat of wax two days later.

QUARRY TILE

Quarry or paver tile, Illus. 97, is ideal for both indoor and outdoor use. Space tiles ¼″, ½″ or ¾″ apart. To estimate spacing, place one row the full length and width of area to be covered. Select the size joint that lessens the need to cut tiles. Next make a layout stick. Mark position of each tile and exact spacing on a 1x2. Set tiles in a mortar mix using premix, or 1 part cement, ½ part hydrated lime to 4 parts clean sand. Add water to make a plastic mix. Spread mortar over area you can conveniently reach. Dust wet mortar with a thin coating of dry cement. Trowel in lightly. Mix a small batch of cement and water to the consistency of thick cream. Paint this cream on back of each tile before pressing in position. Starting at outside corner, place each tile according to layout stick. Press firmly in mortar. Check with level and straight edge. Keep all tiles square with straight edge.

After tiles have been allowed to set two or more days, grout joints. Use prepared grout, or mix 1 part white Portland cement to four parts of fine sand. Clean grout off tiles immediately. Finish joints flush or with concave jointer. Wipe tiles clean with water.

When used on front steps, porch or entrance hall, paver tiles add a designer's touch to your home. Matching bullnose paver tile can be used to finish edge of step treads and risers.

CONCRETE TREE REPAIRS

If you need to make a major surgical operation on a tree trunk, Illus. 98, consider these steps. Using an adze or curved chisel, remove as much decay as possible. Remove fungus. Fungus will only continue to destroy the tree. Use special care not to harm the growing layer just under the bark. Protect the edges of the opening by painting with prepared tree paint or shellac. This will keep the edge from drying out.

After removing interior decay, paint the interior with creosote or other solution your nursery recommends for destroying fungus. Do not apply this on the edge of the opening where it can come in contact with the growing cells under the bark.

If you are covering a large area, after applying creosote, apply a thick coating of hot tar. Make it real thick. This not only acts as a sealant, but also as an expansion joint.

If you are filling a cavity greater than 6″ deep, cut 6x6 reinforcing wire to shape of cavity and place these horizontally, 3″ up from bottom, in position 1″ from back of cavity.

Using a fairly dry mix consisting of one part cement to three parts of sand, pack the concrete into bottom of cavity. Fill the cavity 6″ at a time. Be sure to tamp it into every crevice. Leave no air holes. Use a wetter mix around reinforcing wire.

To curve the filling to shape of trunk, make a form, Illus. 99. Using a jig or keyhole saw, cut 1x8 by length needed to curve of trunk. Nail a 6″ wide strip of galvanized metal in position indicated.

Drill holes in board. Tie a length of clothes line to board in position shown, Illus. 99. Tie form in position required.

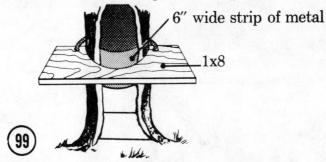

6″ wide strip of metal

1x8

99

As previously mentioned, pour 6″ sections. Shape face to curve of trunk.

Cut a piece of #15 felt and lay on top of each 6″ section. Pour next 6″ section. After packing concrete 3″, lay reinforcing wire in place and continue to fill up to 6″ height. Always lay felt over each section. Always cut felt short so it finishes about ½″ from front. Use a groover to finish edge of joint.

Always fill cavity to inner edge of growing bark. This allows bark to grow over edge of concrete.

TREE BENCH

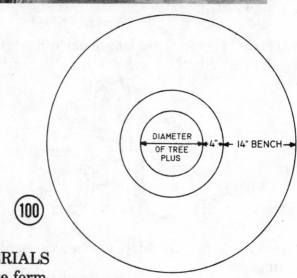

DIAMETER OF TREE PLUS | 4" | 14" BENCH

(100)

LIST OF MATERIALS

10' — 6" concrete form
1 — ¾ x 4 x 8' exterior plywood
1 — 2 x 4 x 4'
1 — 1 x 2 x 4'
4 — ½ x 6" bolts, nuts and washers
Cement, sand, gravel or premix
20' — ⅜ x ½" reinforcing rod
36 — 1¼" lag screws
16 — 2" lag screws

A tree bench, Illus. 100, must allow for at least a 10 year growth. Estimate present diameter of tree. Ascertain normal species growth per year, multiply by ten, add 4" for clearance.

Cut A, Illus. 101, to length required. Use A and B to draw outline.

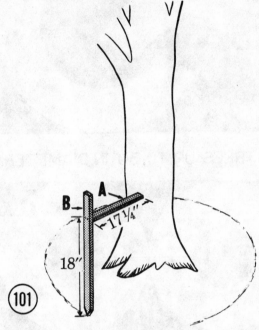

For a 14" wide bench, draw a notch in a piece of 1 x 2 to indicate distance from trunk in ten years, plus 4". Drive a nail in position. Use this to draw inside circle of bench. Add 14" for outside edge of bench, Illus. 102.

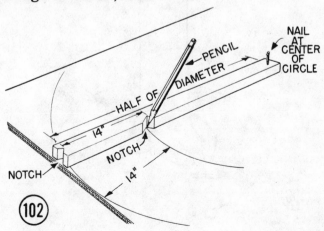

Cut four half circles from ¾" exterior grade plywood. Place one in position so you can position four equally spaced posts, Illus. 103.

FOR TREES UP TO 30" IN DIAMETER

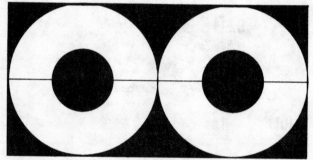

FOR TREES UP TO 12" IN DIAMETER

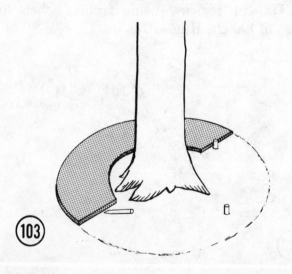

You can drive 4 x 4" posts sharpened at one end into ground, or dig holes at least 16" deep in position forms require. Posts should be plumb and level at 14" height. To insure these are equal height and level, set up guide lines. Check line with line level. Check forms to make certain each is plumb and level. Backfill to hold form in place. Drive two ½ or ⅝" reinforcing rods in position shown, Illus. 104.

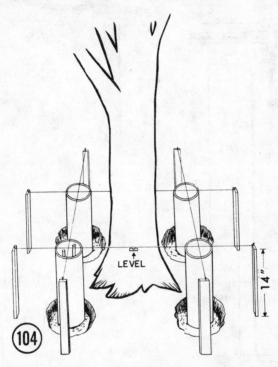

Fill form with a wet mixture of one part cement, two parts sand to three parts gravel, or use a premix. Pour concrete to within 6" from top. Puddle concrete to eliminate any air holes. Use a sharpened 1 x 2.

Drill a ½" hole through center of a paint stirrer. Place bolt in center of form. Screw nut on bolt so 1¼" of thread projects above ¼" thick stirrer. Complete filling form. Remove stirrer. Bolt should project 1½" above cement.

Drill 1" holes ½" deep in center of 2 x 4 x 12", Illus. 105. Then drill ½" hole through 2 x 4. Paint with wood preservative. When dry apply one or two coats of exterior paint.

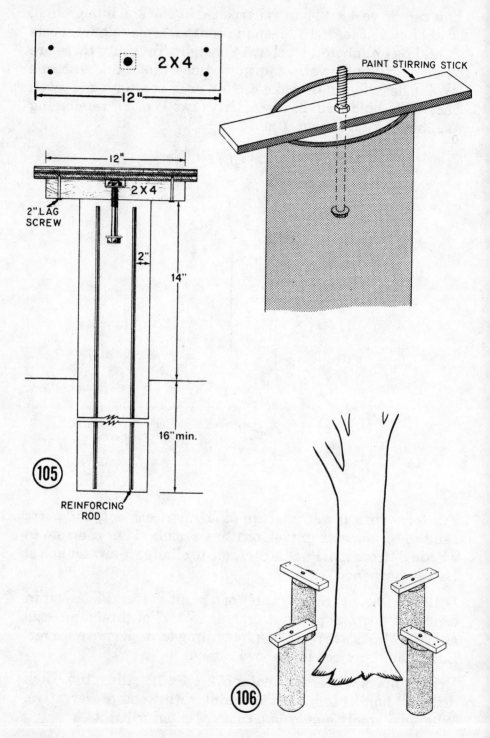

2 X 4

12"

12"

2 X 4

2" LAG
SCREW

2"

14"

16" min.

(105)

REINFORCING
ROD

PAINT STIRRING STICK

(106)

Drill four equally spaced ¼" holes through 2 x 4, Illus. 105. Allow concrete to set three days. Strip form. Fasten 2 x 4 to post, Illus. 106.

Fasten single thickness seat section in place with 2" lag screws. Drill holes through lower section. Apply glue and clamp second section, Illus. 107, in position. Be sure to overlap joint in lower section. Fasten lower section to upper with 1¼" lag screws, 1" from where joints overlap.

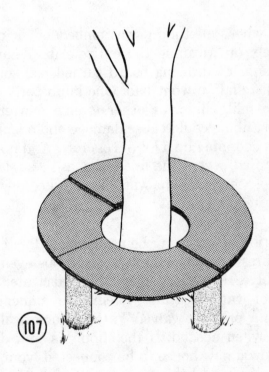

PLEASE NOTE — 4 x 4 wood posts can be used in place of concrete posts. Fasten 2 x 4 seat brace to post with ⅜ x 4" lag screw.

REPAIRING HOT WATER LINE
BURIED IN A CONCRETE SLAB

In the late forties and through the fifties, many builders installed baseboard hot water heating, as well as slab radiation. The latter consisted of burying copper tubing in concrete floor slabs. From an engineering standpoint, both systems looked like great ideas.

The concept of baseboard radiation served by copper tubing buried in concrete and thus protected from damage was sound since the waste heat would warm the concrete. But few installers reckoned with time and repair.

Slab radiation, where all tubing was embedded in concrete, depended entirely on warming the entire slab. What looked great on the designer's drawing board turned real sour as installation aftr installation went bad. One improperly soldered connection, or a nail, cinder or other foreign matter, buried alongside copper tubing could cause damage and a leak no one can find without chopping up a lot of concrete. And no one can tell, once it is buried, where the line leaks, as most slabs were laid above a dirt or sand fill. The water first leaked down, filled a pocket, then spread upward.

The first indication of anything being wrong is loss of water, Illus. 108. The pressure gauge continually drops even though the automatic valve, Illus. 109, continually feeds water into the system. Most hot water heating systems are connected to a supply line with a pressure valve. Water automatically flows into the system when needed. If the supply is shut down for repairs, or to allow a new house to be connected, your heating unit could drain itself, and the pressure gauge would drop accordingly.

In a hilly neighborhood, houses on the highest points could, in times of low water pressure, be subjected to a syphoning action. Water under pressure that should flow into the system through a pressure valve designed to work only in one way, would allow

74

water to be withdrawn if the valve became clogged with foreign matter, sediment, etc.

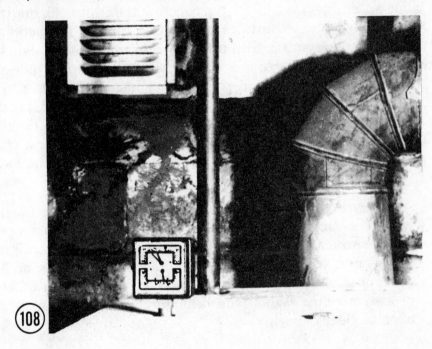

(108)

(109)

When baseboard radiation is installed in rooms above a basement, finding a leak presents few problems. In houses where hot water lines are buried in a concrete slab, Illus. 40, finding a leak invariably necessitates taking up carpeting, finished flooring or underlayment, then breaking into the concrete slab. Unless the homeowner knows where each line is buried, unless he has an accurate plan of the original installation, he can be robbed blind. Locating a line, then a leak is quite a feat. It takes patience and a willingness to proceed very slowly. Unless great caution is taken, you can drive a concrete breaker or a chisel into copper tubing and cut a copper line even before you locate where its buried.

As every homeowner who lives with a leaky roof soon learns, water flows many mysterious and sneaky ways. When a pipe in a slab springs a leak, unless you are very rich, don't call in a plumber or heating contractor. Go first to your doctor and borrow a stethoscope. Ask him to show you how to use it, then probe your floor. When you hear gurgling, you begin to zero in on the leaky spot.

As many floor slabs contain 2x3 or 2x4 sleepers, A, Illus. 40, and wire reinforcing B, you will need an electric hammer, a hand saw and wire cutters. Rent the electric hammer and wire cutters.

Use extreme care when using an electric hammer. Wear safety goggles and only use the hammer after your dealer checks you out. An electric hammer is an easy tool for an intelligent, alert person to use, but a dangerous tool for anyone who doesn't recognize its power.

Always start a hole some distance away from where you think the pipe is located. Check your heating system to make certain a feeder line isn't located in area selected. A feeder line is one that supplies each radiator. If you can possibly locate the contractor who made the installation, he can save you considerable time, labor, mental anguish and money. If you can't, get first hand advice as to where pipes are buried. Proceed very cautiously until you break open a hole in concrete. Use every care possible to break this hole clear of any pipe. Using a hammer and chisel, chip hole larger until you can get your hand under slab. Dig out earth and keep feeling around until you find the pipe. Always use a hammer and chisel when you work near the pipe. Don't gamble using an electric hammer. One touch with an electric hammer and you need to replace more tubing than any leak normally requires.

When you find the leak, open up concrete so a plumber can sweat two nipples and a length of tubing, or do it yourself following directions outlined in Books #675 and #682. If you have to saw through a length of sleeper, it's OK and it won't have to be replaced.

Once you have located the pipe, make a chart showing distance from wall, its direction and depth below floor. Also indicate position of feeder lines.

Leave repaired pipe exposed until you run a two or three day test. When replacing sand make certain it doesn't contain any nails, etc. Repatch concrete. Replace underlayment, etc., and say a prayer of thanks that you could find and follow directions.

HOW TO INSTALL A SUMP PUMP

A sump pump is usually required to relieve water pressure under floor or around footings. If water collects in one corner of your basement, it can be corrected by digging a hole 12″ in diameter, 12″, 18″ to 24″ deep. Use a chisel and hammer to break through concrete or rent an electric hammer. Since these work fast, rent one by the hour. Cut hole to size and depth sump pump manufacturer recommends. Most manufacturers recommend filling bottom of hole with 3″ to 4″ of gravel, then placing an 8″ to 12″ diameter drain tile vertically in position, Illus. 110. Back fill around tile with gravel, repair floor with concrete. Top of tile should finish flush with floor, then be covered with a plywood cap cut to exact size.

Drill holes in cap for pipe and electric line, several more to allow well to breathe.

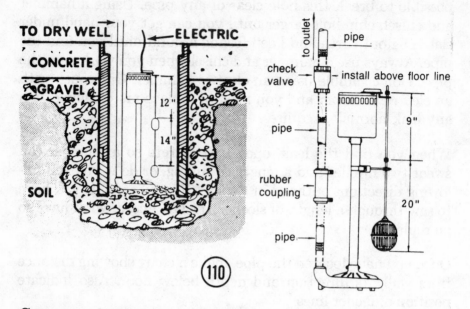

Connect exhaust pipe from pump to dry well. Since you will have to go through foundation wall to discharge water to a dry well, drill a hole through foundation at the highest possible point.

When buying a sump pump, be sure to obtain one that will "lift" water to height your foundation requires.

CAUTION: Never plug in any electrical tool or equipment when you are working in water, or when your hands or feet are wet.

REPAIRING CONCRETE PAVEMENT

First remove all loose and broken pieces. Using a chisel cut away feather edge, Illus. 111, and undercut as noted. Wet edge of existing concrete and paint with a wet mix of cement and water. While still wet, fill area with concrete consisting of 1 part cement, four parts of sand. For large areas, build a form, Illus. 112, and fill with a 1:2¼:3 mix.

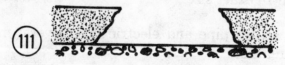

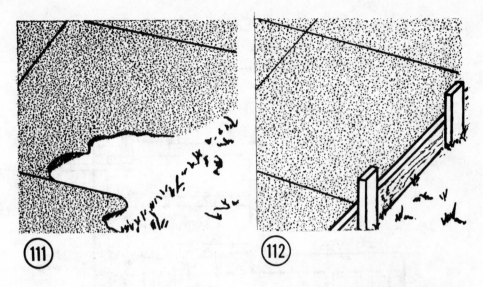

Since a cement and sand mix has a greater tendency to shrink when curing, do this. Use a screed, Illus. 113, with a couple of nail heads in bottom edge. This will permit leveling concrete with a slight rise. When concrete cures it should finish level with existing slab.

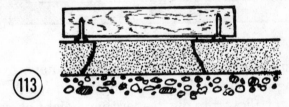

HOW TO BUILD A WATER STORAGE CISTERN

With water in many areas in short supply, a build-it-yourself cistern, Illus. 114, can provide a reserve supply that will carry your garden through a bad drought. It also provides water in case of fire. Locate cistern in a low place, some distance from your house. If your house is on a hillside, locate cistern where excavated earth can be banked around exposed side. While a cistern is a water-tight tank, the overflow could present a problem unless directed away from house.

A cistern can be built any size or shape space permits. Gutters and leaders from house are connected to cistern to catch runoff.

A filtering system of sand and gravel helps catch sediment. While it filters out a lot of the dirt, it still doesn't permit water to be used for cooking or drinking.

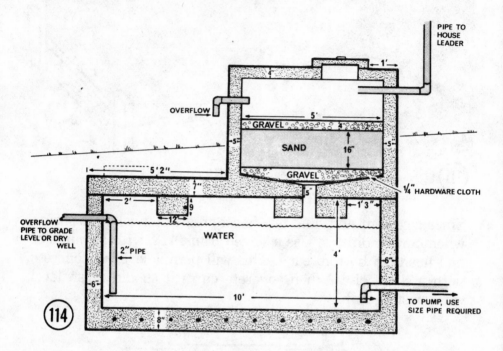

If you are building a new house, arrange to have excavating equipment dig a hole 2 ft. larger in length and width than size of cistern. Don't allow it to dig deeper than height of cistern. Be sure to save all top soil. Pile it some distance from the excavation so it doesn't get mixed with subsoil. Use a pick and shovel to level bottom. Use a straight edge and level to level bottom as accurately as possible. Do not back fill bottom. The cistern should be built on undisturbed soil.

The cistern, connected to house leader, receives rain water filtered through stone, sand and gravel. Cistern water can be connected to a pipe serving a toilet.

Install overflow pipe from filter and tank on side closest to low point of your property. If necessary, empty overflow into a dry well.

80

If additional water is necessary for fire-fighting, locate cistern as far away from house or other buildings as your property permits.

Build form, Illus. 115. Use 2x8's for form for a 3000 or larger size, 2x6's for a 1000 to 2500 gal. tank. Paint lumber with old crankcase oil, nail together at corners, check with square. A form is considered square when diagonals are of equal length.

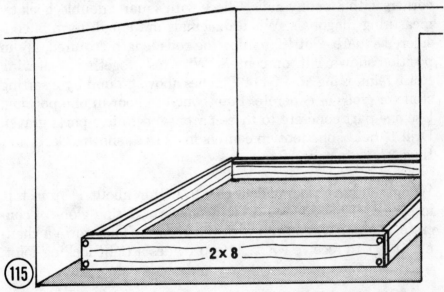

Use ½" reinforcing rods, Illus. 116, to reinforce floor. Fasten together 12" on centers with wire. If you plan on pouring concrete walls, embed ½" reinforcing rods by length required, 12" on centers vertically around perimeter, 3" in from outside edge of floor. Do not embed reinforcing rods vertically if you build sides with concrete blocks.

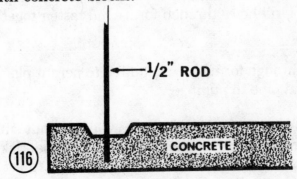

If you install a cistern in an area where rock prevents digging deep, use excavated soil to bank around finished cistern.

Allow 12″ air space between water and top of cistern. In locating cistern, keep water level below frost level.

After leveling bottom and checking same with a straight edge and level, build forms to size required. These can be nailed at corners with 8-penny nails. Check with square; doublecheck by measuring diagonals. When diagonals are equal, form is considered square. Cut ½″ reinforcing rods length required, lay in position shown 1 ft. on centers. Wire rods together with wire. Raise reinforcing about 1 or 2 inches above ground by inserting stone or globs of concrete. Pour concrete floor in one pouring. Use one part concrete to three parts sand to five parts gravel. Drill 1″ holes, one foot on centers in 1x3's as shown. Bevel 45°, Illus. 37. Note page 30.

When you have poured floor slab to within about 1″ from top, insert 1x3, Illus. 37, about 1¾″ from form. Using globs of concrete, position these flush with top of form. They form a channel (key) to receive walls. This provides a tight locking joint,

For a poured concrete wall, embed vertical rods 12″ on centers in position shown, Illus. 116. Allow concrete to begin to set, carefully remove 1x3's. Allow floor to set at least 3 days before building forms for walls. Paint inside face of ½″ plyscord with old crankcase oil. Cut ½″ reinforcing rods to length required to run horizontally, wire in position, Illus. 12. Cut ½″ plyscord for inner form. Drill holes through forms and fasten together with cross ties.

Drill holes through form to size required to permit placing overflow pipe and pipe to pump.

Use ready-mix concrete if available so you can pour entire form in one pouring. Use a broom handle or similar type stick to work concrete down around reinforcing. Allow walls to set at least three days before stripping forms. If necessary, inside face of cistern can be parged with a coat of plaster made from one part cement to two parts fine screened sand. Caulk pipe joints carefully with cement after removing forms.

Build or buy joists and slabs for top, Illus. 117. Your masonry supply house will be able to supply pre-cast slabs of sufficient size to roof over entire top of cistern. If you wish to build a filtering bed, build slabs to size required, allowing a 6-inch opening for filtered water flow.

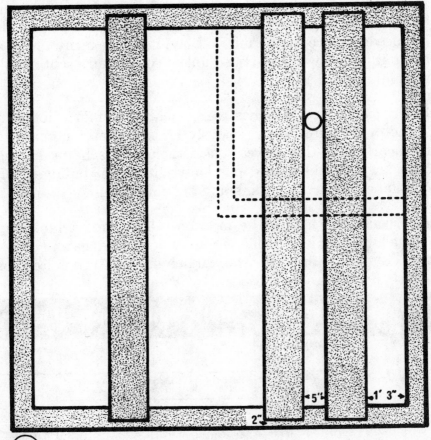

(117)

Build forms for filter bed to 4 ft. height or height space permits. A workable filtering system consists of ¾" gravel placed on top of a ¼" hardware cloth, 12 to 16" of sand, 4" of ¾ gravel on top. Water will run into gravel, seep through sand.

If filter bed is below grade, embed a 4″ drain tile in cap covering filter. This provides an air vent.

Install pump size desired.

FACTS ABOUT A SEPTIC TANK

Before buying property in an undeveloped area, inquire whether the Board of Health will issue a permit for a septic tank. If the property is adjacent to a stream or lake that's part of a watershed, or close to wells supplying a community water system, a permit will only be granted if the tank and field can be placed an approved distance. If the parcel doesn't provide the needed space, no permit will be issued. In many areas, a building permit won't be issued unless you obtain a septic tank permit.

Before signing any purchase agreement, ascertain whether a septic tank permit was previously refused. Smart purchasers frequently save themselves considerable heartache by having their lawyer make a land purchase agreement contingent on obtaining both a septic tank and building permit.

Also consider whether the location selected for a tank and field allows sufficient clear space for a house, garage and driveway. Illus. 118, indicates the various size tanks required for 2, 3, 4 and 5 bedroom houses.

Recommended Septic Tank Capacities

No. of bedrooms in dwelling	Capacity per bedroom in gallons	Required total tank capacity in gallons	Tank Size			
			Inside width	Inside length	Liquid depth	Total depth
2 or less	375	750	3 ft. 6 in.	7 ft. 6 in.	4 ft. 0 in.	5 ft. 0 in.
3	300	900	3 ft. 6 in.	8 ft. 0 in.	4 ft. 6 in.	5 ft. 6 in.
4	250	1,000	4 ft. 0 in.	8 ft. 0 in.	4 ft. 6 in.	5 ft. 6 in.
5	250	1,250	4 ft. 0 in.	9 ft. 0 in.	4 ft. 6 in.	5 ft. 6 in.

(118) See page 92 for new method of laying concrete blocks.

Prospective property buyers who plan on building outside of a sewer district, should study the installation of a septic tank and field, as it's important in making the property usable. Most rural communities adjacent to any size city, only permit installation if it's placed 5' or more from the house, at least 50' from a cistern, 100' from a well, and that the field is not closer than 10' from property line, Illus. 119.

When selecting a site be sure to consider wells, cisterns, or streams on your neighbor's property.

As noted in Illus. 119, 120, codes specify hub and spigot, or hubless cast iron sewer pipe between house and septic tank. This must be 5'0" or longer. Cast iron or bituminous sewer pipe is acceptable between the septic tank and either a junction or distribution box.

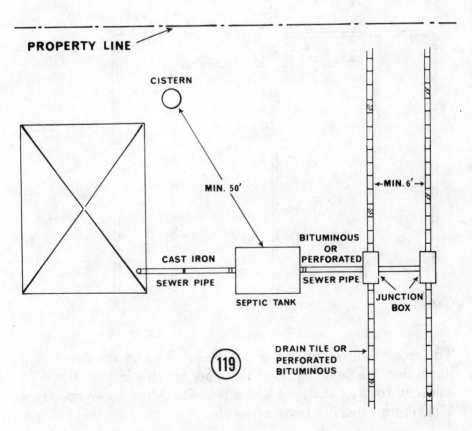

PROPERTY LINE

CISTERN

MIN. 50'

←MIN. 6'→

BITUMINOUS OR PERFORATED

CAST IRON

SEWER PIPE

SEWER PIPE

SEPTIC TANK

JUNCTION BOX

119

DRAIN TILE OR → PERFORATED BITUMINOUS

The sewer line between house and septic tank must slope at least ¼″ to 1″ per foot, as does the connection between the tank and the junction box. Follow local code requirements. The septic tank must be level.

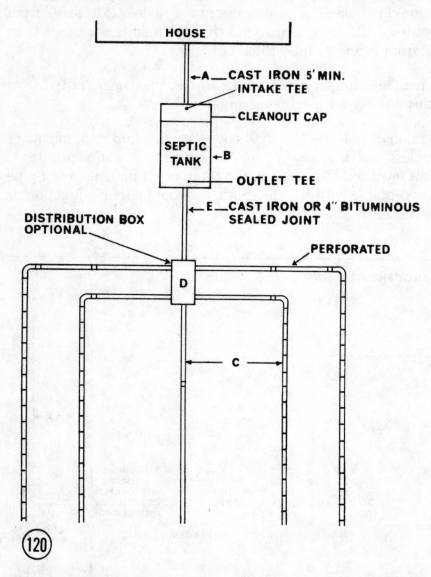

The area required for the field is determined by the number of lineal feet the field requires. You can run this in two, three, or as many rows as shape of space permits. Most codes specify a 6′ minimum spacing between fields.

A septic tank and field should never be located in a low, wet area, or on rock, and only in an open area where no tree roots, shrubbery, etc., will send roots down to the tile. After selecting a site, call the Department of Health and they will make a free inspection or will advise what must be done to pass inspection.

Most inspectors not only inspect the site, make inquiry concerning the number of bedrooms the house will contain, recommend size of tank, shape, and length of field, but will also make an absorption test to ascertain how fast the subsoil will absorb water. Or they will advise you how to make a test.

While installing a system is relatively simple, a septic tank installation can become real sticky when you learn that such and such a septic tank company can obtain a permit for you, then quotes an exorbitant cost.

Regardless of what an "in" septic tank company may tell you, no one can stop you from making your own installation, providing you follow local regulations.

If you think anyone in the health department is giving you a hard time, there are two good reasons why this may be so. Most health departments make an honest and determined effort to protect the area adjacent to water shed property. Regardless of whether it's a nearby stream, runoff into a stream, or a well on your property, or neighboring property, the field must be laid according to regulations. The second reason is financial. If a local septic company has a monopoly, they may overcharge.

If you read and learn how to install a septic tank before you start talking business to an installer, you will save yourself a lot of time, money and aggravation. If you then decide to do part or all the work, you can save even more money. Either way, you get a better job when you know what needs to be done.

A septic tank system consists of a large metal or concrete holding tank B, Illus. 120; a distribution center D; and a field of 4″

diameter perforated drain tile, C. Many codes permit installation without a distribution box. They allow junction boxes as shown in Illus. 119.

The sewer line E, Illus. 120, must be 4″ hub and spigot or hubless cast iron, 5′0″ or more in length. Hubless 4″ cast iron, now available in many areas, is also approved by most health departments.

You can buy a metal septic tank or a prefabricated concrete one, or build a 6″ form, and pour the concrete tank in place. Reinforce sides and bottom with ½″ rods, or use 6x6 reinforcing wire.

Since tanks purchased readymade will be delivered by a truck with a crane that positions the tank level in the excavation you have prepared, the question you must resolve is whether the site selected for the tank can be reached by a truck. Give consideration to this simple but important fact since once you start excavating for a house you might not be able to truck and crane a tank into position.

If you prefer to pour a tank in place, will a ready mix truck be able to come close enough to chute or wheelbarrow the ready mix? Or does the location of the site automatically require you to rent a concrete mixing machine and do the work on the site? With planks properly placed to form a level, or downhill walk, two or three willing souls can usually wheelbarrow concrete from a ready mix truck fast enough to pour a tank without paying overtime for a waiting truck. If you decide to buy ready mix, ask the company how long you can hold the truck on the job before overtime sets in.

Always check tank with a level after it's placed in position, and before the truck leaves.

Trenches for a field are usually 18″ wide and to a depth that permits 4″ to 6″ of gravel on bottom, 4″ tile, plus 2″ to 4″ of gravel on top, Illus. 121. The 4″ drain tile used in the field must be placed at a pitch codes specify. This could be 2″ to 4″ in 100′.

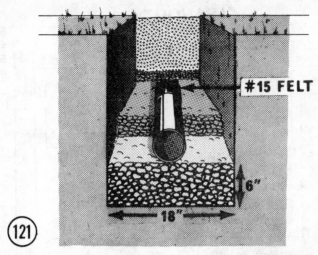

#15 FELT

6"

18"

(121)

Building a form and pouring a tank is only recommended when you have alienated the affection of local suppliers.

If you buy a prefab, buy the size your house requires, or one size larger. Buy a size that would be required if you add extra bedrooms at some date in the future. There is no economy in putting in a minimum approved size when a larger size only costs a few dollars more.

If you buy a complete installation, the installer will dig the hole for the tank and trench the field. If you decide to dig it, be sure to excavate to exact depth the Health Department suggests, and make certain bottom is level. Use a straight 2x4 and a level to check bottom of excavation. The depth and size of hole for a tank is determined by its size and shape.

Since the sewer line from the house to the tank must pitch ¼″ per foot, or pitch local codes require, the depth of the hole will be determined by the intake TEE, Illus. 122. Dig trench from house to selected site. Slope trench to pitch pipe requires. When 5′0″ or further away, place a TEE, Illus. 123, in position on the end of the pipe. To estimate depth of excavation, measure from bottom of TEE to depth below TEE your tank requires. Note position of inlet TEE, Illus. 122.

Codes recommend outlet TEE be placed 3″ lower than inlet.

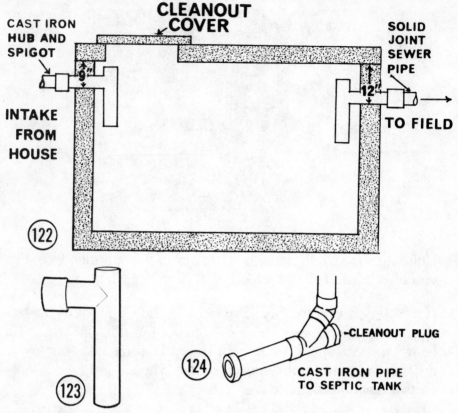

CLEANOUT COVER

CAST IRON HUB AND SPIGOT

9"

INTAKE FROM HOUSE

SOLID JOINT SEWER PIPE

12"

TO FIELD

122

123

124

CLEANOUT PLUG

CAST IRON PIPE TO SEPTIC TANK

When installing a new tank, always install a cleanout plug, Illus. 124, in direct line with the inlet TEE. This greatly simplifies servicing a septic tank.

When installing a septic tank on rolling or hilly land the field should follow the contour of the land.

A septic tank can also be built with concrete blocks. Excavate to size and depth tank requires. Level bottom, then dig a 10" wide footing trench, 4" lower than floor, around perimeter of floor area. Lay a thick bed of mortar and position a starter course of 10x8x16 or 18" blocks. These will project 4" above the floor area.

Cut 6x6 reinforcing wire to size floor area requires. Raise wire about 1" and pour floor flush with top of block. Fill cores of blocks with concrete. Allow slab to set three days, then build walls using 8x8x16 or 18″ blocks. These are placed flush with inside edge of 10″ block. Fill cores of the first three courses of blocks with concrete. If you are building a large tank, lay reinforcing wire, Illus. 66, in every third course of block.

Set Inlet Tee and Outlet Tee in position shown, Illus. 122.

Allow tank to set at least five days, then paint inside and outside with hot tar or asphalt cement.

Follow local code requirements, and lay field to depth specified. Cover with size and depth of gravel specified, then cover gravel with #15 felt before backfilling. Since the Board of Health will want to make an inspection before you cover the field, follow directions they provide, and under no circumstances give the inspector a hard time.

Cover tank with precast slabs, Illus. 125.

When laying perforated bituminous field tile, follow local code recommendations regarding a distribution box, Illus. 120, or junction boxes, Illus. 119.

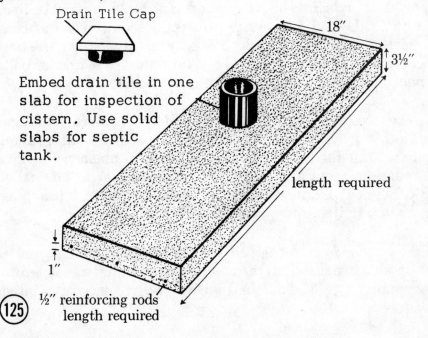

Drain Tile Cap

Embed drain tile in one slab for inspection of cistern. Use solid slabs for septic tank.

18″

3½″

length required

1″

(125) ½″ reinforcing rods length required

NEW METHOD SIMPLIFIES
LAYING CONCRETE BLOCK

Once or twice in every lifetime, those alert to change witness an economic, social or industrial happening that alters the lifestyle for millions. Those who appreciate what is happening get lucky fast. As this book goes on press, a new, easy, fast and practical method of laying concrete block has been tested and code approved. It can turn amateurs into pros on their first attempt.

No longer do those wanting to build a house, addition, basement entry, retirement home, or any building project that requires concrete blocks, have to depend on a skilled mason. Everyone who can lift and position a block alongside a taut line can now lay up their own foundation, do a professional job and save a bundle.

While directions starting on page 42 explain the conventional method of laying block in which the ends of each block are buttered up with mortar and each course laid in a bed of mortar, the new system permits laying blocks up dry. No mortar at end joints or between courses. Only the first course, Illus. 59, is laid according to directions outlined. In areas subject to high winds, codes not only specify locking the first course to footings in a bed of mortar, but also to anchor bolts spaced eight feet apart, or distance specified by codes.

All courses above first course are laid dry, no mortar. Stagger all joints, overlap blocks on corners, Illus. 62,63. After laying blocks, the inside and outside surface is plastered with a Portland cement and glass fiber prepared mortar mix called BlocBond. This only requires adding clean water. It is applied ⅛" thick to the entire surface, including first course.

This method not only locks all blocks together, but also makes the wall more resistant to moisture. BlocBond comes in white, gray and beige. The finished surface can be left as is, painted, stuccoed, etc.

92

While this new method is by far the easiest, care must be used in setting corner blocks square, each course level, all courses plumb.

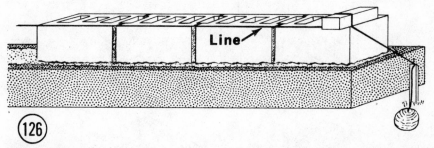

Line

(126)

Use a guide line, Illus. 59, 126. Lay each block to the line. Continually check each course with a level. Check each corner with a square. As previously mentioned, lay the first course in mortar. Make certain this is level and corners square. If you need to anchor first course to footing, embed 12 or 15" bolts and washers, Illus. 127, in footing in position cores in block permit. Space anchor bolts every 8'. Allow bolts to project 7" above footing. Anchor these bolts to blocks by filling cores in first course.

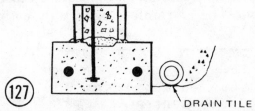

(127)

DRAIN TILE

Start by positioning corner blocks. Lay blocks along the entire wall, corner to corner, or corner to a window or door. Overlap joints at corners and on each course. Do not use reinforcing between courses. Use corner blocks, Illus. 63, at a door opening.

Use special channel blocks, Illus. 128, for a steel or aluminum window. Use blocks, Illus. 129, for a wood sash.

As in conventional foundation work, those using 8 x 8 x 16" blocks should lay a 16" wide footing. This can be poured into a form made from 2 x 4, Illus. 58, or where codes or conditions require, use 2 x 6.

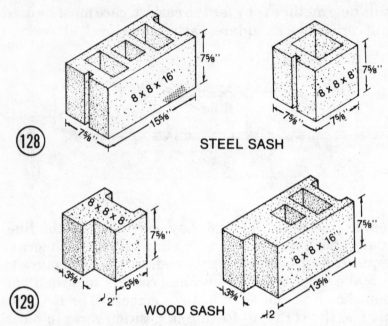

(128)

STEEL SASH

(129)

WOOD SASH

10 x 8 x 16 or 18" blocks should be laid over a 20" wide footing. Use 2 x 6 for footing forms. Those building a foundation for a full basement, and/or a one story concrete block structure, should embed two ½ or ⅝" reinforcing rods in footings, Illus. 130.

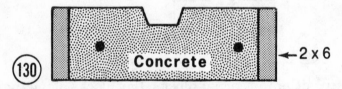

(130)

Concrete

←2 x 6

Lay corner blocks in a bed of mortar, Illus. 58. To locate exact corner, drop plumb bob down from guide lines, Illus. 58. Position a corner block, Illus. 131, square, level and plumb in corner.

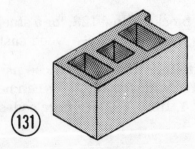

(131)

Stretch guide line between corner blocks, Illus. 59.

Lay first course of block for outside walls and partitions in a bed of mortar. Butter up end of each block, Illus. 61. Lock partition blocks into every other course of outside wall, Illus. 132.

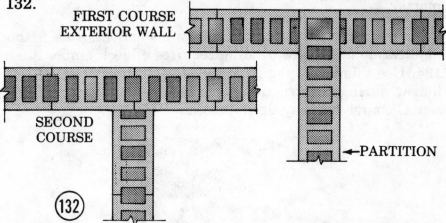

FIRST COURSE
EXTERIOR WALL

SECOND
COURSE

←PARTITION

(132)

Fill core in blocks that receive anchor bolts, Illus. 133.

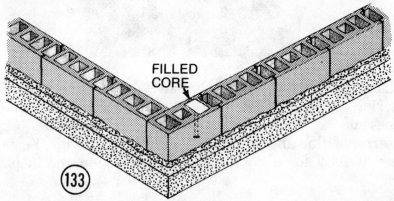

FILLED
CORE

(133)

Go completely around the foundation laying blocks dry. While pros will lay up a full basement wall before bonding inside and outside surfaces, a beginner should lay up only as many courses as he can surface bond in a day's time. This permits locking all blocks laid, and allows the surface to harden overnight. It also permits washing out the mixing tub as quickly as possible to eliminate any build up.

Follow the manufacturer's directions concerning the use life of each batch of mortar. Wash out your mixing tub as quickly as you have used up a batch. Always dip your trowel in a clean pail of water to prevent any build up of mortar.

When you need to cut a block, try to make a square cut. If you have difficulty, fill any broken edge with surface bonding mortar.

Besides a clean mixing tub, pool trowel and float, Illus. 5, the application of surface bonding requires a block guide, Illus. 134. Use this when you have to leave a course overnight or longer during construction. This insures joining each horizontal course halfway up the block.

You can make a block guide. For an 8" block, cut a piece of ¾" exterior grade plywood A, Illus. 135, 7⅝" x 4'0". Cut two pieces of ⅛" tempered hardboard B, 3¾" x 4'0". Apply waterproof glue and nail B to A with 4 penny nails. Fasten a piece of 1 x 2 for a handle or buy and fasten one in place.

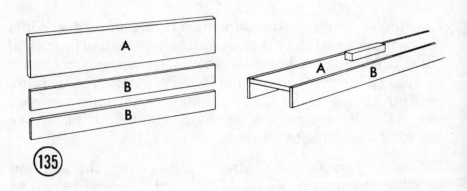

BlocBond should cover one half of the top course during construction. Where you need to make a vertical joint at the end of a day's work, use a piece of ⅛ x 6" x length required hardboard, Illus. 136. Butt the mortar a full ⅛" thick so it covers one half of each block. Use care not to apply mortar more than ⅛" thick. Slide your block guide along top course to be surfaced so you make a neat joint.

HARDBOARD

Allow 48 hours before backfilling. Use caution when backfilling a 7' or 8' basement wall. Always shovel fill in close to wall. Keep mechanical equipment away from wall.

Those building a basement should brace a long basement wall every 8', Illus. 137, for at least 48 hours in normal weather. Leave braces in place until all backfilling is completed. If you plan on going into business, hinge 2 x 6 brace.

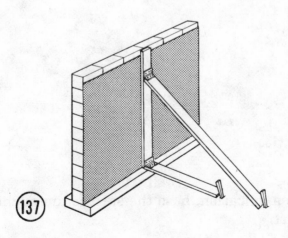

BlocBond comes in 70 or 80 lb. bags. While now being shipped into selected sales areas, those having any building plans should make inquiry as to when same will be available. It's worth waiting for. You will find it real easy to use.

Always read and follow manufacturer's directions. While directions suggest laying up a complete foundation, we suggest the beginner lay up only as many courses required to provide 100 or more square feet of surface.

A 70 lb. bag will cover 100 square feet. This could be 50 square feet on both sides of a wall, or 100 on one side. To more accurately estimate square foot surface area for dry stacked blocks, refer to Concrete Block sizes on pages 124, 125, also Illus. 139.

Pour 3 gallons of clean water into a clean mixing tub, Illus. 5. Add and stir a bag of BlocBond. Stir for 2 or 3 minutes, not more. What you want is a pliable mortar mix. Too much mixing and it could get lumpy. Start applying immediately. Use a mortar board and a pool trowel, Illus. 138. Spread BlocBond ⅛" thick over the entire surface, including the bottom course of block laid in mortar. Smooth the mortar over entire surface with an even pressure but don't force it or keep working over any one spot.

(138)

Just prior to application, hose the surface thoroughly so all blocks are wet.

Actual Height of Units, Inches	No. of Units per 100 sq. ft.	Approximate Bags per 100 sq. ft. of Wall*
7 5/8 (Modular)	121	2.5
3 5/8 (Modular)	255	2.5
8 (Nonmodular)	115	2.5

*Includes application to both sides of 100 sq. ft. wall, based on 70-pound bag. Actual coverage depends upon the waste, the application and the texture of the masonry unit.

Always follow this procedure:

1. Wet surface thoroughly before applying BlocBond.

2. To join each application, cover only half the surface of adjacent blocks, Illus. 134.

3. BlocBond must be applied within an hour and a half after mixing.

4. It's essential to wash out the mixing tub and clean tools as quickly as possible to prevent mortar from hardening.

Test using BlocBond on a small job. Build a block planter, fence, etc. After you get dry behind the ears, you will be able to estimate how much of a mix you can handle, how long it takes to apply, how to plan a day's work. If you are building an addition, porch, patio or garage, or enclosing your property with a low or high concrete block fence, you can either mix large batches of BlocBond and cover a much wider area, or stack more blocks after you have completed surfacing the blocks laid.

Those building a full basement and those who turn professional can rent a cement mixer. Only do this if you have extra help. Rent a mixer with rubber paddles. Have an assistant hose out the mixer after making each batch to keep it sparkling clean.

As you apply BlocBond, do not exceed ⅛" thickness. Press firmly but do not apply pressure. A smooth or stucco effect, Illus. 140, can be achieved with a wide heel trowel, Illus. 5.

The stipple effect, Illus. 141, can be obtained with a mason's float, Illus. 5.

A swirled effect, Illus. 142, can be done with a hard brush.

After allowing BlocBond to set, following manufacturer's directions, the surface can be painted or plastered. Use latex based paints indoors, an acrylic type latex for exterior application.

Since all blocks have to be thoroughly hosed down before applying BlocBond, and only a light spray can be applied while a previously surfaced area is setting up, how you schedule your work is important. Allow BlocBond to set overnight before hosing dry blocks above. Do not apply BlocBond when the surface will be exposed to immediate rain. If time is short, buy a roll of polyethylene and cover the wall after you finish a day's work. Secure the polyethylene with blocks at top and at bottom, Illus. 143. Keep the polyethylene away from wall so it doesn't bond to surface. Give the wall plenty of air so it can dry overnight.

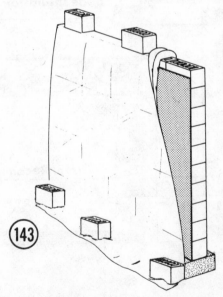

Cores in the top course of block should be partially filled with rumpled paper, then covered with mortar and troweled smooth. Set all anchor bolts or plate fasteners in top course, Illus. 144, 145, in mortar filled cores following directions outlined. Those building a one story concrete block house, addition, porch, etc., must position plate or rafter fasteners where plans specify. These lock the plate and rafters to wall.

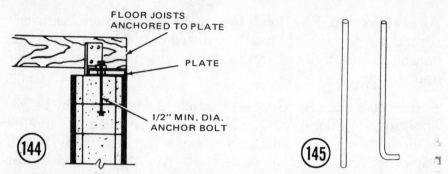

FLOOR JOISTS
ANCHORED TO PLATE

PLATE

1/2" MIN. DIA.
ANCHOR BOLT

(144)

(145)

Applying BlocBond only requires using the proper trowel or float. Smoothing the exterior surface permits bonding rigid insulation, Illus. 144, 146, to inside and/or outside where climate requires same.

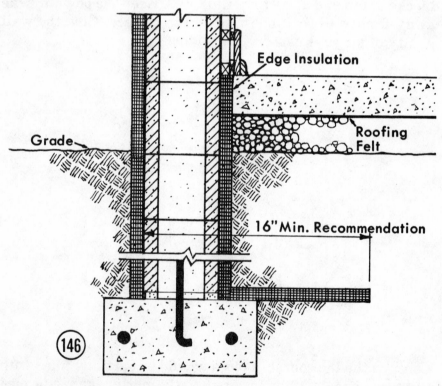

Edge Insulation

Roofing Felt

Grade

16" Min. Recommendation

(146)

BlocBond now permits those desiring to build their own house to raise four walls with concrete block. The savings in labor costs is sizeable, plus being tax free. In addition, a concrete block house, with a series of smaller and higher windows not only saves heat, but also offers greater protection.

WINDOW INSTALLATION

Due to the wide variation in design and size of steel, aluminum and wood windows, currently being manufactured, directions offered only provide a guide. It's important to follow directions the window manufacturer specifies concerning size of opening, framing, nailing, or use of anchoring fasteners. Always allow exact amount of space manufacturer or retailer recommends for calking. Install flashing exactly as manufacturer suggests.

Those installing a window or door in a block wall must build the wall up to height of lintel over window, Illus. 147.

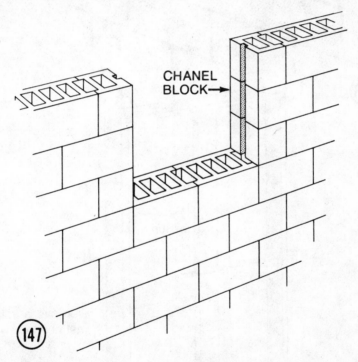

CHANEL
BLOCK→

(147)

Channel blocks, Illus. 148, designed to receive a steel or aluminum sash, are positioned where window requires same. Windows can be slid down from the top and temporarily supported at height required. Use a piece of brick, stone or block to hold window at exact height. Insert putty, wedges or fastening manufacturer of window recommends. Trowel a sill in place. Bevel sill to outside.

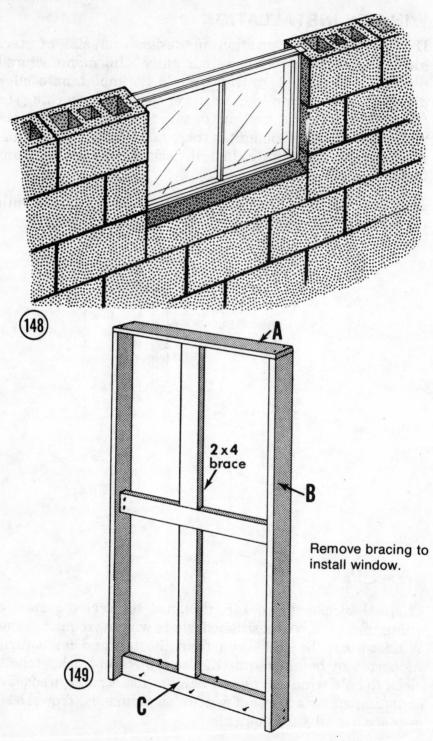

148

149

A

B

2 x 4
brace

Remove bracing to
install window.

C

Where a window requires a buck, Illus. 149, build it to size window manufacturer suggests. A buck simplifies pouring sill after setting window in place. Position buck in opening so it finishes flush at top with a course of blocks, Illus. 150. To build a buck, nail A to B. Nail B to C. Use 2 x 4 or size lumber window manufacturer recommends for C. Drive some 1" big head nails into side and top of C.

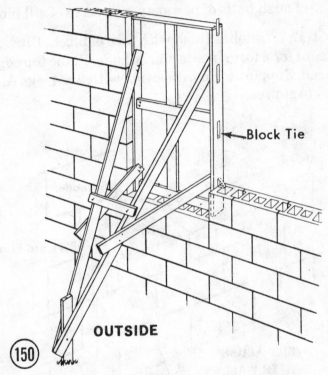

←—Block Tie

OUTSIDE

(150)

Nail block ties to B, Illus. 151, at height needed so each can be bent between blocks, Illus. 150.

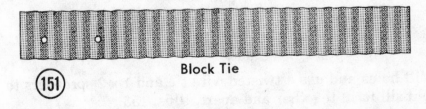

Block Tie

(151)

Use finished end blocks to build wall up to top of window. Square up buck, level and plumb.

Allow ¼", or spacing window manufacturer suggests, between block and buck. Bend ties into each or every other course of block. Fill joint between block and buck with non-hardening calking.

Knock out temporary bracing and fasten window to buck. Test open window.

Rumple up and push balls of newspaper in cores of sill blocks.

To build a sill that finishes flush with face of block, Illus. 152, use ¾" plyscord for a form inside and outside. Plane top edge to angle required. Support form in position with 2 x 4 legs A, and angle braces to stakes.

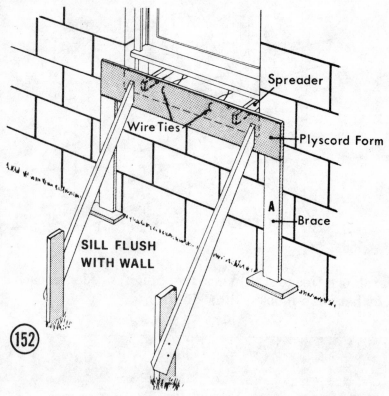

Drill holes and use a twisted wire tie and 1 x 2 spreaders to hold sill form together and apart, Illus. 153.

Mix one part Portland cement to three parts sand for a mortar mix. Use color cement that matches blocks. Remove spreader as you fill form. Work mortar up under window sill and pack it in. Trowel finished sill to pitch shown. Allow sill to set at least four days. Cut wire and remove form. Recut wire, plaster over end.

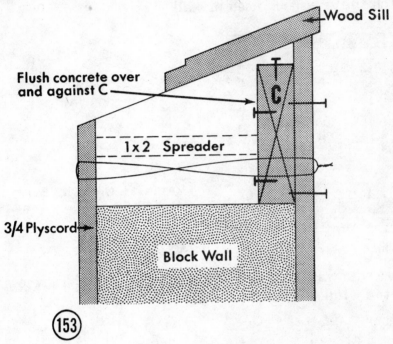

Wood Sill

Flush concrete over and against C

C

1 x 2 Spreader

3/4 Plyscord

Block Wall

(153)

If you want the sill to project over face of block, place a 1 x 4 or thickness needed to project amount desired, Illus. 154. Nail plyscord form to back edge of buck. Don't drive nails all the way. Brace plyscord form in position, Illus. 155. Nails rest on sill.

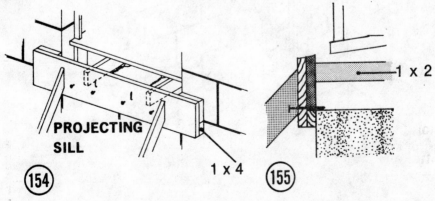

PROJECTING SILL

(154)

1 x 4

1 x 2

(155)

107

After windows have been blocked in, you can either buy a precast lintel, Illus. 156, and ask the concrete products dealer to hoist it in position when he delivers same; or use lintel blocks, Illus. 157. Lintel blocks are available in width that match the size block used in wall.

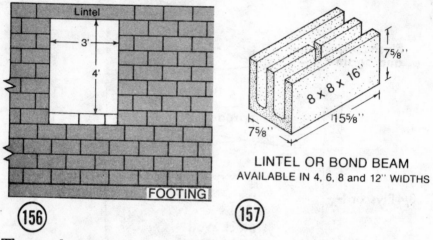

LINTEL OR BOND BEAM
AVAILABLE IN 4, 6, 8 and 12" WIDTHS

(156)   (157)

Those who want to build a lintel in place should fill recess in front of window with 2 x 4 shoring, Illus. 158.

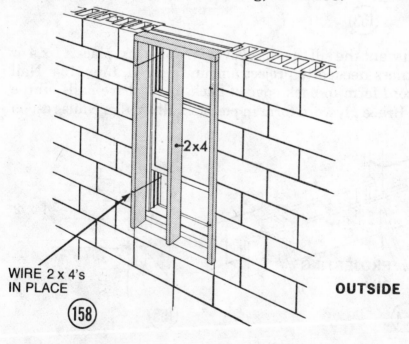

WIRE 2 x 4's
IN PLACE

OUTSIDE

(158)

Place lintel blocks, groove side up, end to end, Illus. 159. Fill grooves with 1" of concrete. Cut ½" reinforcing rod full length of assembled blocks and embed same in position. Fill grooves with another 2" of mortar and embed two more rods. Fill block to top. Allow beam to set undisturbed for four or more days, then complete perimeter course of block. Apply BlocBond surfacing.

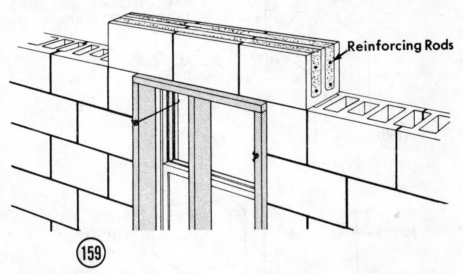

Reinforcing Rods

(159)

Calk joint around windows with calking window manufacturer recommends.

ANCHOR PLATE

Set anchor bolts or plate fasteners, Illus. 160. Fill cores of blocks that contain anchor fasteners with mortar. If codes specify same, fill other blocks with rumpled newspaper, then cover with 1 or 2" of mortar.

Where codes require a poured perimeter beam, those laying up blocks for a one story house, addition, garage, etc., must build a form, Illus. 161. Your concrete products retailer sells hangers that hold rods in position required. They also sell 18-gauge galvanized steel ceiling joists or truss anchors, Illus. 162. These are positioned before pouring beam. This simplifies installing ceiling joists or truss rafters.

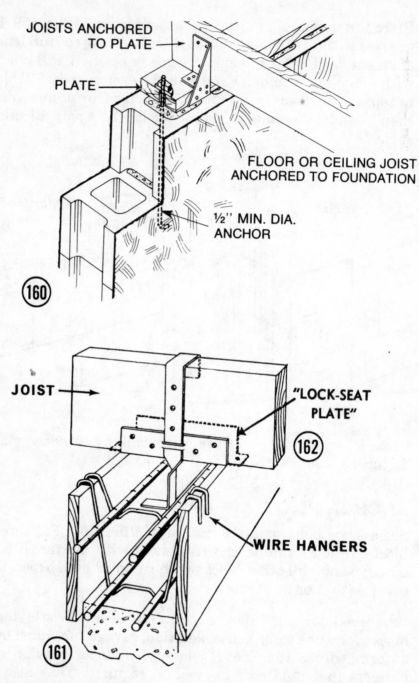

JOISTS ANCHORED TO PLATE

PLATE

FLOOR OR CEILING JOIST ANCHORED TO FOUNDATION

½" MIN. DIA. ANCHOR

160

JOIST

"LOCK-SEAT PLATE"

162

WIRE HANGERS

161

Form for poured perimeter beam. Note hangers for reinforcing rods; truss anchors for ceiling or floor joists. These are also used to anchor prefabricated rafters.

CRAWL SPACE VENTILATION

Illus. 163 shows one type of metal ventilator that can be installed in the top course of block used to ventilate crawl space. Normally two of these, one at each end, is sufficient. Where you have a moisture problem, or where crawl space requires free circulation of air, use two additional ones equally spaced on a long wall. Or use basement windows. In areas that experience severe winters, buy vents that can be closed in cold weather.

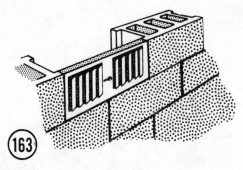

Wire screening bent over edge of a block turned on its side, Illus. 164, can also be used as a vent. Use fine wire mesh to discourage bees.

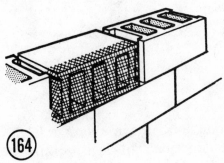

FACTS TO REMEMBER

Where a block wall is left exposed in space used for living, use bullnose sash blocks, Illus. 165.

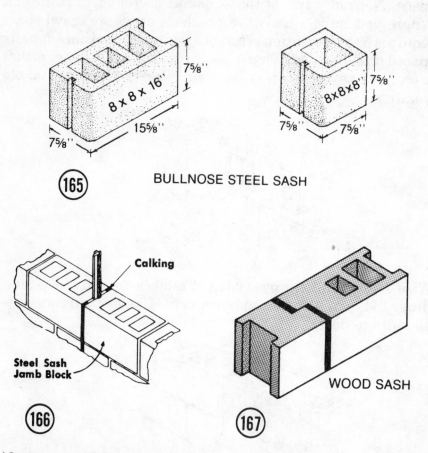

7⅝"

8 x 8 x 16"

15⅝"

7⅝"

8x8x8" 7⅝"

7⅝"   7⅝"

(165) BULLNOSE STEEL SASH

Calking

Steel Sash
Jamb Block

(166)

WOOD SASH

(167)

Always allow an expansion joint in a long or high wall, Illus. 166. These relieve contraction. This is especially important on walls over 30' in length. One method of relieving pressure is through the use of steel sash blocks positioned in a vertical line. These accommodate a preformed rubber control joint, Illus. 168. Install a control joint where building specification and/or code require.

The wood sash block, Illus. 167, can also be used as an expansion joint. Allow ½", or width gap height and/or length of wall require. Fill gap with calking.

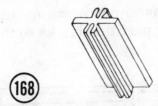

(168)

PREFORMED RUBBER CONTROL JOINT

Expansion joints are important between a floor slab and footing or foundation, Illus. 169. Also between a floor slab and a footing for a steel basement column.

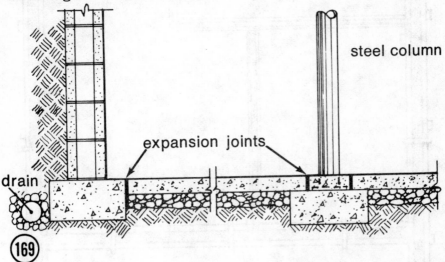

steel column

expansion joints

drain

(169)

Use a chisel to cut a block. Draw a line clear around block. Place block on a level surface or on a bag of sand, Illus. 170. Make cuts an even depth all along drawn line rather than attempt to cut through. Always stagger joints so cut blocks don't line up.

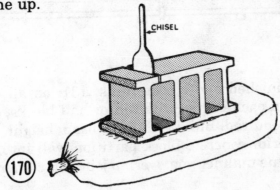

CHISEL

(170)

113

Build and pour footing forms for partitions and exterior walls at one time, Illus. 171. Lay partition blocks so they lock into exterior walls, Illus. 132.

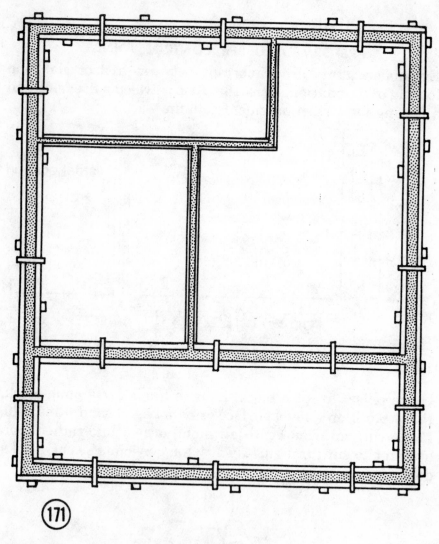

(171)

Always use finished end blocks, Illus. 131, on all exposed corners. Fill in each course with full or half blocks, or cut a block to size required. Build wall up to beam height. Always lay in an interior concrete block partition wall into outside wall in the same manner you overlap joints in corner blocks, Illus. 63.

Where 2 x 4 or 2 x 6 framing is used for basement partitions, anchor sleeper clips, Illus. 172, to footings. Install close to ends of each partition, also every eight feet. The anchor is nailed to the shoe with 8 penny nails.

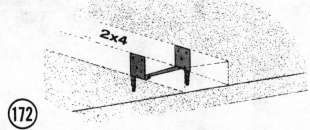

Concrete work in cold weather requires taking special care against freezing. Don't apply BlocBond or lay up blocks in mortar unless temperatures remain above 40° during the time it takes the mortar to cure thoroughly. During cold weather, allow 7 days for BlocBond where it cures within 24 hours in normal temperature.

Always insulate crawl space under a house, Illus. 173.

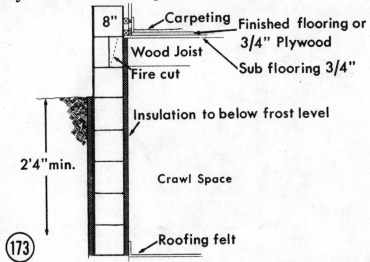

When you apply insulation on outside of foundation, cover with asbestos cement board, or other protective weatherproof panelboard.

Rigid foam insulation not only keeps out cold, but also acts as a compression joint between slab and foundation, Illus. 173.

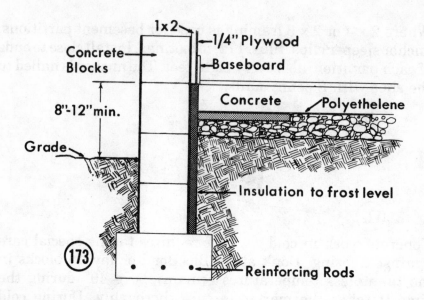

When pouring a slab on grade, bond insulation to foundation, using mastic that insulation manufacturer recommends, Illus. 174, and protect it with asbestos cement board.

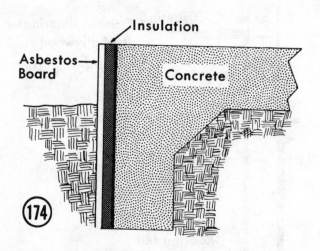

If you find it necessary to break through a BlocBond surfaced foundation to run a fuel, water, electric or sewer line, or if any crack appears due to settling of footing, chip 1" of surfacing away from edge of opening, Illus. 175. Apply a new batch of BlocBond containing special ingredients manufacturer recommends.

On very hot days a well soaked wall may dry out before you have completed surfacing. Use a brush to carefully apply water to dry blocks. Use care so no water penetrates edge of applied BlocBond.

When heat or a high wind starts to dry out applied BlocBond too fast, use a fine spray to add sufficient moisture to slow down the curing process. Only apply a fine spray after BlocBond has cured sufficiently to absorb same.

Those who are building a basement, entry well to a basement door, and those installing doors and windows in a concrete block house, should read Book #697 Forms, Footings, Foundations and Framing. It provides many helpful construction details.

TERMS USED IN CONCRETE WORK

ADDITIVES,Waterproofing,Hardener,AntiFreeze,Dust Inhibitor,page 24
AGGREGATE — refers to sand, gravel or crushed stone
ANCHOR BOLTS — used to fasten framing to foundation, page 47
BASEMENT, WATERPROOFING,page 34
BLOCBONDING — a surface bonding method of laying blocks dry, no mortar, page 92
BLOCK REINFORCING,page 46
BLOCK SIZES, page 124
BRICKLAYING, page 120
CARPETING OVER CONCRETE, page 32
CHALK LINE — used to indicate lines
CHISEL — used in cutting blocks, bricks, page 12,44,113
CISTERN, page 79
COLORS — available dry in powder form to be mixed with dry cement; stain also available for use on finished concrete, page 58
CONCRETE BLOCKS, page 42
CONCRETE, MIXES — A mixture of cement, sand and gravel, page 8
CONCRETE RAKE, page 19
CONCRETE WALKS, page 48
CONTRACTION JOINTS, page 20,112
CONTROL JOINTS, page 20,112
CURING — process of keeping freshly laid concrete covered and wet for period of time, page 13
DIAGONALS, page 62
DRAINAGE TILE — to receive and carry water, page 20
DRY WELL — a stone filled hole large enough to receive water from house drains, page 16
DUST INHIBITOR, page 24
EDGING TOOL — used to round off edges of plastic concrete, page 27
ENTRY DOOR BASEMENT, page 52
EXPANSION JOINT — strip of insulation board.
Permits expansion and contraction of concrete, page 20, 22
FINISHING,page 26
FLOATS — wood or metal, used to smooth surface after concrete stiffens but is still plastic. Wood float leaves gritty, nonskid surface. Metal float finishes smooth, page 26.
FLOORS, page 29
FLOORS OVER WET AREAS, page 31
FOOTINGS, page 29
FORMS — lumber or undisturbed earth used to hold concrete, page 14
GRAVEL — an aggregate ranging in size from ¼" to 1½".
Must be clean and hard when used in concrete.
GROOVER — tool used to groove lines in still plastic concrete, page 21
GROUT — a mixture of cement and sand used to fill joints between bricks, block, flagstone.

ESSENTIALS OF GOOD BRICK CONSTRUCTION

◀ Spread a uniform bed of mortar over only a few brick. Furrow only lightly, if at all. Place plenty of mortar on the end of the brick to be placed. Brick is then shoved into place so that mortar is squeezed out of top of head joint.

◀ After placing, mortar squeezed out of bed joint is cut off to prevent staining the wall.

◀ Concave jointer

◀ When mortar joint becomes thumbprint hard, tool with steel jointer slightly larger
◀ than the mortar joint. Concave or V joints have best weather resistance.

◀ V jointer

◄ When placing closures, place plenty of mortar on ends of brick in place and on ends of brick to be placed. Shove closure into place without disturbing brick on either side.

◄ When a wall is capped with a brick rowlock course, it is essential that all vertical joints be completely filled.

◄ rowlock

In cavity wall construction, mortar droppings should not be permitted to fall into the cavity. An aid in preventing this is to bevel the bed joint ◄ away from the cavity.

BEVELING BED JOINTS

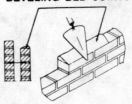

In metal-tied walls, a parge coat between wythes is the barrier to moisture penetration. Parging may be applied to either backup or facing units. In either case, excess mortar squeezed out of the joints should be cut off and ¼ in. to ⅜ in. of mortar trowelled on.

When brick are laid on a beveled bed joint, a minimum of mortar is squeezed out of the joint. Brick (1)—beveled joint; brick (2)—conventional joint.

The mortar squeezed from the joints on the cavity side may be plastered on to the units. This same procedure may be used for laying exterior wythes of reinforced brick walls. Mortar droppings should not be permitted in grout core.

◄ Cavity wall metal ties are embedded in bed joints as units are laid.

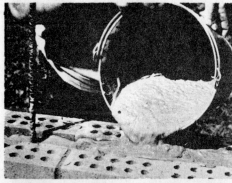

◄ In reinforced masonry, vertical steel should be placed before masonry is laid. Grout should be mixed with sufficient water to cause it to pour readily.

◄ After grout is poured, it should be puddled to consolidate it.

◄ Horizontal steel may be placed in grout core as wall is built.

CONCRETE BLOCK SIZES

A– 7 5/8"   B– 15 5/8"   C–9 5/8"   D– 11 5/8"   E– 3 5/8"

APPROXIMATE METRIC SIZE A-19.37   B-39.7   C-24.45   D-29.53   E-9.21

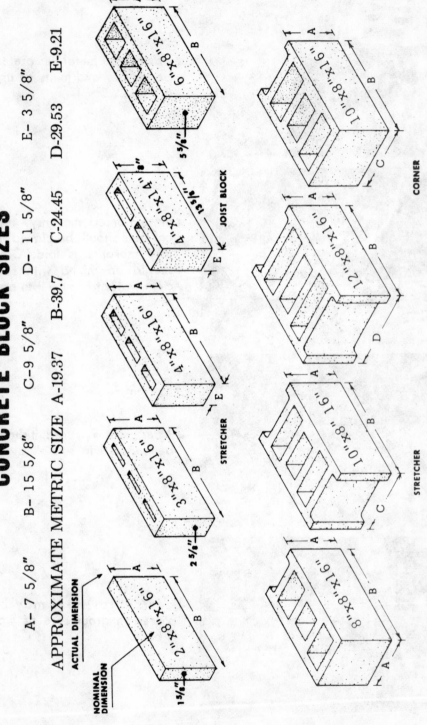

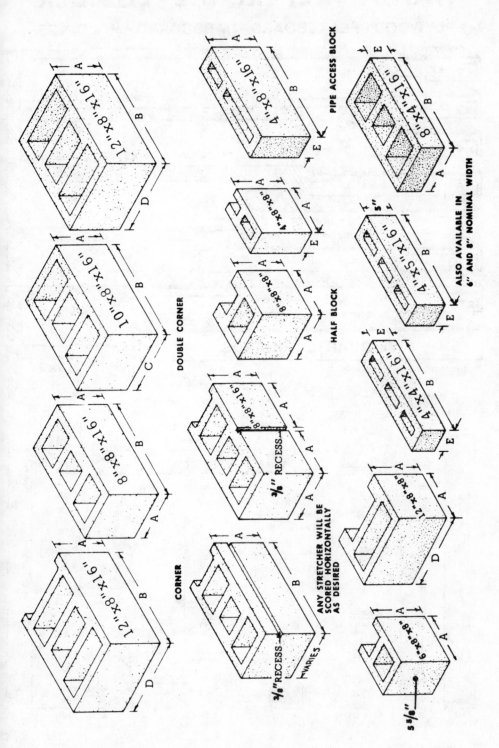

12"x8"x16"

10"x8"x16"

8"x8"x16"

12"x8"x16"

DOUBLE CORNER

CORNER

4"x8"x16"

PIPE ACCESS BLOCK

4"x8"x8"

8"x8"x8"

HALF BLOCK

8"x8"x16"

¾" RECESS

ANY STRETCHER WILL BE
SCORED HORIZONTALLY
AS DESIRED

¾" RECESS

VARIES

4"x4"x16"

ALSO AVAILABLE IN
6" AND 8" NOMINAL WIDTH

4"x5"x16"

5"

4"x4"x16"

12"x8"x8"

6"x8"x8"

5⅝"

125

HANDY - REFERENCE - LUMBER
PLYWOOD - FLAKEBOARD - HARDBOARD - MOLDINGS

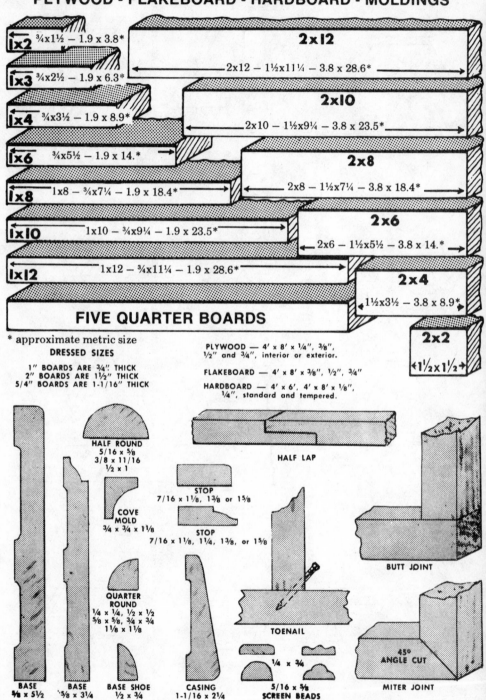

1x2 ¾x1½ — 1.9 x 3.8*

2x12

2x12 — 1½x11¼ — 3.8 x 28.6*

1x3 ¾x2½ — 1.9 x 6.3*

2x10

2x10 — 1½x9¼ — 3.8 x 23.5*

1x4 ¾x3½ — 1.9 x 8.9*

1x6 ¾x5½ — 1.9 x 14.*

2x8

2x8 — 1½x7¼ — 3.8 x 18.4*

1x8 1x8 — ¾x7¼ — 1.9 x 18.4*

2x6

1x10 1x10 — ¾x9¼ — 1.9 x 23.5*

2x6 — 1½x5½ — 3.8 x 14.*

1x12 1x12 — ¾x11¼ — 1.9 x 28.6*

2x4

1½x3½ — 3.8 x 8.9*

FIVE QUARTER BOARDS

2x2

1½x1½

* approximate metric size

DRESSED SIZES

1" BOARDS ARE ¾" THICK
2" BOARDS ARE 1½" THICK
5/4" BOARDS ARE 1-1/16" THICK

PLYWOOD — 4' x 8' x ¼", ⅜",
½" and ¾", interior or exterior.

FLAKEBOARD — 4' x 8' x ⅜", ½", ¾"

HARDBOARD — 4' x 6', 4' x 8' x ⅛",
¼", standard and tempered.

HALF ROUND
5/16 x ⅝
3/8 x 11/16
½ x 1

HALF LAP

STOP
7/16 x 1⅛, 1⅜ or 1⅝

COVE MOLD
¾ x ¾ x 1⅛

STOP
7/16 x 1⅛, 1¼, 1⅜, or 1⅝

QUARTER ROUND
¼ x ¼, ½ x ½
⅝ x ⅝, ¾ x ¾
1⅛ x 1⅛

BUTT JOINT

TOENAIL

BASE
⅝ x 5½

BASE
⅝ x 3¼

BASE SHOE
½ x ¾

CASING
1-1/16 x 2¼

¼ x ¾

**5/16 x ⅝
SCREEN BEADS**

**45°
ANGLE CUT**

MITER JOINT

HANDY REFERENCE-NAILS

Common — Finishing —

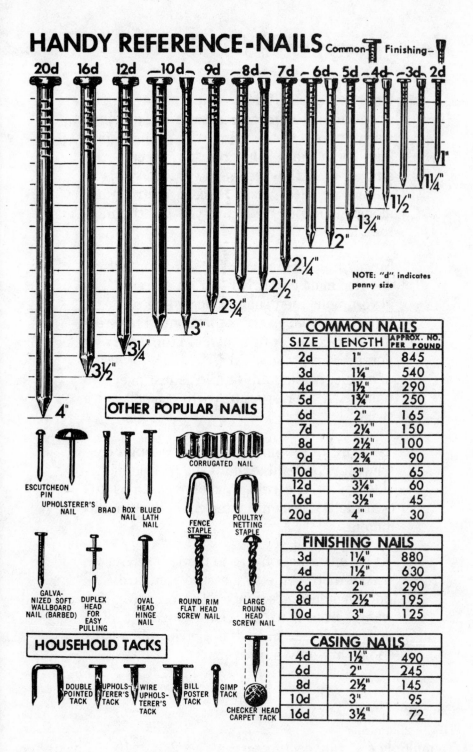

20d 16d 12d —10d— 9d —8d— 7d —6d— 5d—4d— —3d— 2d

1"

1¼"

1½"

1¾"

2"

2¼"

2½"

2¾"

3"

3¼"

3½"

4"

NOTE: "d" indicates penny size

OTHER POPULAR NAILS

ESCUTCHEON PIN

UPHOLSTERER'S NAIL

BRAD

BOX NAIL

BLUED LATH NAIL

CORRUGATED NAIL

FENCE STAPLE

POULTRY NETTING STAPLE

GALVA-NIZED SOFT WALLBOARD NAIL (BARBED)

DUPLEX HEAD FOR EASY PULLING

OVAL HEAD HINGE NAIL

ROUND RIM FLAT HEAD SCREW NAIL

LARGE ROUND HEAD SCREW NAIL

HOUSEHOLD TACKS

DOUBLE POINTED TACK

UPHOLS-TERER'S TACK

WIRE UPHOLS-TERER'S TACK

BILL POSTER TACK

GIMP TACK

CHECKER HEAD CARPET TACK

COMMON NAILS

SIZE	LENGTH	APPROX. NO. PER POUND
2d	1"	845
3d	1¼"	540
4d	1½"	290
5d	1¾"	250
6d	2"	165
7d	2¼"	150
8d	2½"	100
9d	2¾"	90
10d	3"	65
12d	3¼"	60
16d	3½"	45
20d	4"	30

FINISHING NAILS

3d	1¼"	880
4d	1½"	630
6d	2"	290
8d	2½"	195
10d	3"	125

CASING NAILS

4d	1½"	490
6d	2"	245
8d	2½"	145
10d	3"	95
16d	3½"	72

127

HOW TO THINK METRIC

Government officials concerned with the adoption of the metric system are quick to warn anyone from attempting to make precise conversions. One quickly accepts this advice when they begin to convert yards to meters or vice versa. Place a metric ruler alongside a foot ruler and you get the message fast.

Since a meter equals 1.09361 yards, or 39⅜″ +, the decimals can drive you up a creek. The government men suggest accepting a rough, rather than an exact equivalent. They recommend considering a meter in the same way you presently use a yard. A kilometer as 0.6 of a mile. A kilogram or kilo as just over two pounds. A liter, a quart, with a small extra swig.

To more fully appreciate why a rough conversion is preferable, note the 6″ rule alongside the metric rule. A meter contains 100 centimeters. A centimeter contains 10 millimeters.

As an introduction to the metric system, we used a metric rule to measure standard U.S. building materials. Since a 1x2 measures anywheres from ¾ to $^{25}/_{32}$ x 1½″, which is typical of U.S. lumber sizes, the metric equivalents shown are only approximate.

Consider 1″ equal to 2.54 centimeters;
10″ = 25.4 cm.
To multiply 4¼″ into centimeters: 4.25 × 2.54 = 10.795 or 10.8 cm.

EASY-TO-USE-METRIC SCALE

DECIMAL EQUIVALENTS

Fraction		Decimal
1/32		.03125
	1/16	.0625
3/32		.09375
	1/8	.125
5/32		.15625
	3/16	.1875
7/32		.21875
	1/4	.250
9/32		.28125
	5/16	.3125
11/32		.34375
	3/8	.375
13/32		.40625
	7/16	.4375
15/32		.46875
	1/2	.500
17/32		.53125
	9/16	.5625
19/32		.59375
	5/8	.625
21/32		.65625
	11/16	.6875
23/32		.71875
	3/4	.750
25/32		.78125
	13/16	.8125
27/32		.84375
	7/8	.875
29/32		.90625
	15/16	.9375
31/32		.96875

FRACTIONS — CENTIMETERS

Fraction		Centimeters
1/16		0.16
	1/8	0.32
3/16		0.48
	1/4	0.64
5/16		0.79
	3/8	0.95
7/16		1.11
	1/2	1.27
9/16		1.43
	5/8	1.59
11/16		1.75
	3/4	1.91

Fraction		Centimeters
13/16		2.06
	7/8	2.22
15/16		2.38

INCHES — CENTIMETERS

Inches			Centimeters
1			2.54
	1/8		2.9
		1/4	3.2
	3/8		3.5
		1/2	3.8
	5/8		4.1
		3/4	4.4
	7/8		4.8
2			5.1
	1/8		5.4
		1/4	5.7
	3/8		6.0
		1/2	6.4
	5/8		6.7
		3/4	7.0
	7/8		7.3
3			7.6
	1/8		7.9
		1/4	8.3
	3/8		8.6
		1/2	8.9
	5/8		9.2
		3/4	9.5
	7/8		9.8
4			10.2
	1/8		10.5
		1/4	10.8
	3/8		11.1
		1/2	11.4
	5/8		11.7
		3/4	12.1
	7/8		12.4
5			12.7
	1/8		13.0
		1/4	13.3
	3/8		13.7
		1/2	14.0
	5/8		14.3
		3/4	14.6
	7/8		14.9

			cm
6			15.2
	1/8		15.6
		1/4	15.9
	3/8		16.2
		1/2	16.5
	5/8		16.8
		3/4	17.1
	7/8		17.5
7			17.8
	1/8		18.1
		1/4	18.4
	3/8		18.7
		1/2	19.1
	5/8		19.4
		3/4	19.7
	7/8		20.0
8			20.3
	1/8		20.6
		1/4	21.0
	3/8		21.3
		1/2	21.6
	5/8		21.9
		3/4	22.2
	7/8		22.5
9			22.9
	1/8		23.2
		1/4	23.5
	3/8		23.8
		1/2	24.1
	5/8		24.4
		3/4	24.8
	7/8		25.1
10			25.4
	1/8		25.7
		1/4	26.0
	3/8		26.4
		1/2	26.7
	5/8		27.0
		3/4	27.3
	7/8		27.6
11			27.9
	1/8		28.3
		1/4	28.6
	3/8		28.9
		1/2	29.2
	5/8		29.5
		3/4	29.8
	7/8		30.2

			cm
12			30.5
	1/8		30.8
		1/4	31.1
	3/8		31.4
		1/2	31.8
	5/8		32.1
		3/4	32.4
	7/8		32.7
14			35.6
16			40.6
20			50.8
30			76.2
40			101.6
50			127.0
60			152.4
70			177.8
80			203.2
90			228.6
100			254.0

FEET	INCHES	CENTIMETERS
1	12	30.5
2	24	61.0
3	36	91.4
4	48	121.9
5	60	152.4
6	72	182.9
7	84	213.4
8	96	243.8
9	108	274.3
10	120	304.8
11	132	335.3
12	144	365.8
13	156	396.2
14	168	426.7
15	180	457.2
16	192	487.7
17	204	518.2
18	216	548.6
19	228	579.1
20	240	609.6